Lecture Notes in Geoinformation and Cartography

Series Editors: William Cartwright, Georg Gartner, Liqiu Meng, Michael P. Peterson

Antoni Moore, Igor Drecki
(Editors)

Geospatial Vision
New Dimensions in Cartography

Selected Papers from
the 4th National Cartographic Conference GeoCart'2008
New Zealand

Springer

Editors

Dr Antoni Moore
School of Surveying
University of Otago
PO Box 56
Dunedin
New Zealand
amoore@surveying.otago.ac.nz

Igor Drecki
School of Geography, Geology
& Environmental Science
The University of Auckland
Private Bag 92019
Auckland
New Zealand
i.drecki@auckland.ac.nz

Additional material to this book can be downloaded from http://extras.springer.com.

ISBN 978-3-642-44705-1 ISBN 978-3-540-70970-1 (eBook)

Lecture Notes in Geoinformation and Cartography ISSN: 1863-2246

Cover images by Tomaž Podobnikar and Jürgen Döllner

Cover design: deblik, Berlin

Printed on acid-free paper

9 8 7 6 5 4 3 2 1

springer.com

Table of Contents

Geovisual Exploration of Uncertainty and Terrain

Web Mapping

About the Authors

Phil Bartie is a PhD candidate at the Geospatial Research Centre, University of Canterbury, NZ. Previously he has worked in government and commercial sectors designing and implementing GIS solutions. He holds a BSc(Hons) in Human and Physical Geography, and MSc in GIS. His research interests are in location based services, interface design, and visibility analysis.

Contact: *philbartie@gmail.com*

George Benwell is Pro-Vice-Chancellor (Commerce) and Dean of the Business School at the University of Otago, NZ. After working as a surveyor, consultant and University of Melbourne academic, he joined Otago's Information Science Department as a senior lecturer in 1990. He led the Department from 2001-2006 before becoming Dean in 2007. His research and teaching are mainly focused on spatial information processing and analysis, and land and health-related information systems. He currently holds a Bachelor of Surveying from the University of Melbourne, an MPhil from City University, London and a PhD from Melbourne.

Contact: *gbenwell@business.otago.ac.nz*

Grant Carroll graduated from the University of Otago (NZ) in 2006 with a BAppSc in Geographic Information Systems and Software Engineering. He has been working at Marlborough District Council (NZ) as their GIS Technician / Developer since mid 2007. He also does contract work for a property firm specialising in Treaty Settlements. His research interests include the tessellation of space in 2D / 3D and network / route mapping. He is actively working to leverage the most out of the council technology to make it easier and more accessible for all users.

Contact: *grant.carroll@marlborough.govt.nz*

William Cartwright is President of the International Cartographic Association. He is Professor of Cartography and Geographical Visualization in the School of Mathematical and Geospatial Sciences at RMIT University, Australia. He holds a Doctor of Philosophy from the University of Melbourne and a Doctor of Education from RMIT University. He has six other university qualifications - in the fields of cartography, applied science, education, media studies, information and communication technology and graphic design. He joined the University after spending a number of years in both the government and private sectors of the mapping industry. His major research interest is the application of integrated media to cartography and the exploration of different metaphorical approaches to the depiction of geographical information.
Contact: *w.cartwright@rmit.edu.au*

Juliane Cron studied cartography at the HTW Dresden and finished her studies in 2006 with a diploma thesis about "Graphical User Interfaces of Interactive Atlases". Since November 2006 she works at the Institute of Cartography at the ETH Zurich in the project "Swiss World Atlas interactive" and is responsible for the editorial concept, i.e. usability, design and implementation of the GUI.
Contact: *juliane.cron@karto.baug.ethz.ch*

Jürgen Döllner is a Professor at the Hasso-Plattner-Institut of the University of Potsdam, directs the computer graphics and visualisation division. He has studied mathematics and computer science and received a Ph.D. in computer science. He researches and teaches in real-time computer graphics, spatial visualisation, software visualisation, and spatial data infrastructures..
Contact: *doellner@hpi.uni-potsdam.de*

Igor Drecki is currently a manager of the Geo-graphics Unit at the School of Geography, Geology and Environmental Science, The University of Auckland, New Zealand. His previous professional experience included private, government and academic environments where he worked as a cartographer. Igor's research interests focus on geographical information uncertainty representation, and also include cartographic design and education. He is currently the Editor of ICA News, the newsletter of the International Cartographic Association, and Vice President of the New Zealand Cartographic Society.
Contact: *i.drecki@auckland.ac.nz*

Igor Florinsky is a Senior Research Scientist at the Institute of Mathematical Problems of Biology, Russian Academy of Sciences, where he has been since 1993. He completed his PhD in that year at the Moscow Institute of Engineers for Geodesy, Aerophotography, and Cartography (MIIGAiK). His research interests include: development of algorithms, methods, and software for digital terrain modelling; multi-scale studies of soil and landscape processes; problems of structural geology; interaction between environment and human activities (using methods of digital terrain modelling, GIS, and spatial statistics); study of the temporal domain in soil, geophysical, and biological processes by singular spectrum analysis; and influence of the geological environment on human health and behaviour. In 1998-2004, he worked at the Manitoba Land Resource Unit, Agriculture and Agri-Food Canada, and Department of Soil Science, University of Manitoba, Winnipeg.
Contact: *iflorinsky@yahoo.ca*

Lorenz Hurni is Professor of Cartography and director of the Institute of Cartography at the ETH Zurich since November 1996. Under his lead, the multimedia "Atlas of Switzerland" as well as a new interactive version of the "Swiss World Atlas", the official Swiss school atlas, are being developed. The emphasis of his research lies in cartographic data models and tools for the production of printed and multimedia maps. Another focus of research covers interactive, multidimensional multimedia map representations. The new possibilities are being explored in international, interdisciplinary projects and being imparted to a broad audience in lectures and courses for students and practitioners.
Contact: *hurni@karto.baug.ethz.ch*

Markus Jobst is a Post Doctoral research grant fellow at the Hasso-Plattner-Institut, University of Potsdam. Besides his commercial multimedia-cartographic activities and the coordination of the scientific projects for photogrammetric documentation and cartographic heritage, Markus has substantial knowledge of digital photography, reproduction and multimedia dissemination processes. His main foci in scientific work are in communication of spatial related data, multimedia 3D cartography, digital cartographic presentation methods including crossmedia publishing and the creation of management tools with the help of digital cartography.
Contact: *markus.jobst@hpi.uni-potsdam.de*

Julian Kardos completed his PhD on the visualisation of uncertainty in Geographic Information Systems in 2005. Since completing his PhD, Julian has been active in GIS research and publications. Julian continues his involvement in the GIS community undertaking work with Intergraph Corporation, a GIS solution provider, as a Business Development Manager and Consultant. Julian has extended his focus beyond the visualization of uncertainty into collaboration and spatial data infrastructures.

Contact: *julian.kardos@intergraph.com*

Simon Kingham is an Associate Professor in the Department of Geography, at the University of Canterbury (Te Whare Wananga O Waitaha). He came to the University of Canterbury in 2000 having completed his PhD at Lancaster University and having worked at the Universities of Newcastle, Huddersfield and Hertfordshire in the UK. His research focuses on urban, issues including transport and accessibility usually with a GIS framework.

Contact: *simon.kingham@canterbury.ac.nz*

Peter Knight is a lecturer in Hydrographic Surveying at the School of Surveying, University of Otago, New Zealand. His research interests include Common Pool Resource Management and he has been studying the social, commercial and spatial aspects of the Bluff oyster fishing community in southern NZ. Peter graduated in 1980 with a Diploma in Hydrographic Surveying from Humber College in Toronto. This start in surveying enabled him to begin a 10 year university career that produced a BA (Guelph 1987); BSc (Surveying-Toronto 1990); and an MASc (Civil Engineering-Toronto 1994). Peter's career as a Hydrographer continued with his appointment as a multi-disciplinary hydrographer with the Canadian Hydrographic Service (1993), a position he held until coming to Otago. He came to New Zealand from Canada in 1997.

Contact: *peter.knight@surveying.otago.ac.nz*

Jan Eric Kyprianidis is a Ph.D. student at the Hasso-Plattner-Insitut of the University of Potsdam. His research interests include non-photorealistic rendering and digital image processing. He received a diploma in mathematics from the University of Hamburg, Germany. Before joining HPI, he worked as a senior software engineer at Adobe Systems.

Contact: *kyprianidis@hpi.uni-potsdam.de*

Adrian Miles is Senior Lecturer in Cinema and New Media in the School of Applied Communication, RMIT University, Melbourne. He is well known for this research in hypertext, specifically the relation of hypertext to cinema and link poetics. He is known internationally for his innovative practice based research in the field of online video, work which he commenced in 2000. He is widely recognised as a pioneer of videoblogging as an aesthetic and critical practice, and as the world's frist videoblogger. The concept of 'softvideo', which Miles introduced in 2003, has been used by several other theoreticians in their own practice, is regularly used in several syllabi nationally and provides a paradigmatic shift in the conception of time based media online.

Contact: *adrian.miles@rmit.edu.au*

Steven Mills completed his PhD in the Department of Computer Science, University of Otago, in 2000. After a year working in industry he accepted a lectureship in computer science at the University of Nottingham. In 2006 he left Nottingham to take up the role of senior research scientist at the newly formed Geospatial Research Centre in Christchurch, NZ. Dr Mills' research interests lie in computer vision and image processing, with a particular focus on motion analysis, 3D scene reconstruction, and applications linked to navigation and georeferenced imagery.

Contact: *steven.mills@grcnz.com*

Antoni Moore has just been appointed as a senior lecturer in Geographic Information Science at the School of Surveying, University of Otago, New Zealand, having previously been in their Department of Information Science since 2001 and before that as a coastal / marine GIS Analyst at Plymouth Marine Laboratory in the UK. His research interests include geovisualisation, which encompasses the visualisation of uncertainty, cognitive mapping and application of virtual / augmented reality. Other research interests cover cartographic generalisation, spatial data structures and use of GIS-related technology in a decision support context (as well as using visualisation for this purpose, intelligent information systems – the subject of his PhD, grassroots mapping and spatio-temporal modelling have been applied).

Contact: *amoore@surveying.otago.ac.nz*

Brian Morris, PhD, has published extensively - in Australia and internationally - in his field of urban and cultural studies on the mediatized nature of cities. Together these publications foreground his concern with bringing into dialogue the transdisciplinary insights of spatial theory, cultural studies, and technology and communication studies as they pertain to the analysis of particular geographical locations. Morris' essays have appeared in leading international and national journals and in books. Morris'more recent research on the Tokyo area of Shibuya has focused on the theorising of relations between experiences of urban places and uses of communication technologies (such as mobile telephones) by users and how these articulate with older practices of navigation.

Contact: *brian.morris@rmit.edu.au*

Tomaž Podobnikar currently works at the Scientific Research Centre of the Slovenian Academy for Sciences and Arts. He holds a BSc in geographical information systems and its applications in environment and archaeology, MSc in Monte Carlo methods and its applications in geographical information systems (GIS), PhD in digital terrain modelling (DTM) from various data sources of different quality, all achieved at the University of Ljubljana. He has authored several papers on quality assessments, GIS, DTM and its applications as well as archaeology, biology, dialectology, environment, Mars research, etc. With his own approach he produced a DTM of Slovenia that is currently widely available from the government. He was a research fellow at the University of Technology Delft (The Netherlands), Vienna University of Technology (Austria) and University of Franche-Comté (France). He is currently researching on geomorphological analysis at Institute of the Photogrammetry and Remote Sensing, Vienna University of Technology.

Contact: *tp@zrc-sazu.si*

Holger Regenbrecht has been a senior lecturer in the Department of Information Science at Otago University Since November 2004. He is a computer scientist and has a PhD in engineering from Bauhaus University. He has been working in the fields of Virtual and Augmented Reality for over ten years. He was initiator and manager of the Virtual Reality Laboratory at Bauhaus University Weimar (Germany) and the Mixed Reality Laboratory at DaimlerChrysler Research and Technology (Ulm, Germany). His research interests include Human-Computer Interaction (HCI), (collaborative) Augmented Reality, 3D teleconferencing, psychological aspects of Mixed Reality, and three-dimensional user interfaces (3DUI).

Contact: *holger.regenbrecht@otago.ac.nz*

Nigel Stanger is a lecturer in the Department of Information Science at the University of Otago School of Business, where he has taught in the areas of systems analysis and database systems since 1989. He has active research interests in digital repositories, distributed and web database systems, XML technologies, physical database design and database performance. He was the project lead and programmer for the School of Business EPrints repository, which he continues to maintain and enhance. He is also heavily involved in projects to increase the uptake of digital repository technology within New Zealand, and is a key member of the Open Access Repositories in New Zealand (OARiNZ) project.
Contact: *nstanger@infoscience.otago.ac.nz*

Andrew Ternes graduated from the Royal Melbourne Institute of Technology with a Bachelor of Applied Science in Surveying in 2003. He worked briefly in the land survey industry before joining the Hydrographic Survey department of the Port of Melbourne Corporation (PoMC) in 2004. Andrew was trained in Hydrographic Survey by the PoMC, and encouraged him to gain further knowledge, by supporting his attendance in the Category A Hydrographic Surveying Program at the University of Otago, Dunedin, New Zealand. His involvement in the program, for which he gained a Postgraduate Diploma in 2007, allowed him to explore his own research ideas and interests related to hydrographic mapping and navigation, such as 3D visualisation, real time mapping and augmented reality guided navigation. Elements of these were applied in his dissertation "The implementation of 3D nautical charts to hydrographic survey".
Contact: *andrew.ternes@portofmelbourne.com*

Laurene Vaughan, PhD, is the Director of Research and Innovation in the School of Applied Communication at RMIT University. Laurene is well recognised for her cross-disciplinary approach to design and practice based research. Laurene's emerging contribution to the field of interaction design focuses on understanding the human experience of space and place through technology. She does this through a unique combination of fashion, landscape and human centred design methodologies. Since 2006 she has been actively engaged as a Chief Investigator and Project Leader within the Australasian CRC for Interaction Design (ACID).
Contact: *laurene.vaughan@rmit.edu.au*

Samuel Wiesmann studied geography at the University of Zurich. He finished his studies at the GIS division with a diploma thesis about developing interactive map legends. Since 2007 he works at the Institute of Cartography at the ETH Zurich as a research associate. He is involved in the editorial work of the Atlas of Switzerland.

Contact: *wiesmann@karto.baug.ethz.ch*

Jeremy Yuille is a Senior Lecturer in the School of Applied Communication, RMIT University Melbourne. His research examines the areas of Interaction Design, Generative Systems, and audiovisual Performance, through multidisciplinary collaborative projects in the technology, communications, media and entertainment sectors. Research has been applied to new user interfaces, generative systems for content creation, remote collaboration, media theory, education and musical composition and performance. In 2004 he was appointed Program Manager, Mutiuser Environments, at the Australasian CRC for Interaction Design (ACID), involved in setting an agenda for Interaction Design research specifically looking at environments for remote collaboration and methods for undertaking design research informed by social science methodologies.

Contact: *jeremy.yuille@rmit.edu.au*

Introduction

This book contains selected papers from participants at the 4th National Cartographic Conference GeoCart'2008, held in Auckland, New Zealand in September 2008. It provides a contribution to the literature related to contemporary Geoinformation and Cartography as part of the Springer series "Lecture Notes in Geoinformation and Cartography". The series aims to provide publications that highlight the research and professional activities taking place in this exciting discipline area. Books published thus far cover a wide range of topics and their content reflects the diverse nature of interests of contributors in the field.

The GeoCart conferences are held every two years and attract attendees from Australasia and globally. They offer a forum for reflecting on past practices, exploring future possibilities and reporting on the findings of research undertakings. They make valuable contributions to the theory and praxis of Geoinformation and Cartography.

The editors of this book, Antoni Moore, from the University of Otago, and Igor Drecki, from the University of Auckland, have provided contributions that fall under the categories of representation, egocentric mapping, the exploration of tangible and intangible geographical phenomena by visual means and Web mapping. The chapters provide valuable information from contributors that illustrate the exciting developments in the discipline.

I applaud the efforts of the editors and authors for providing this work as an insight into their fields of activity. I hope that you find this book, from the land of the Long White Cloud, a valuable resource.

William Cartwright

School of Mathematical and Geospatial Science
RMIT University, Australia

Preface

This book contains a set of papers selected from research presented at the 4th National Cartographic Conference GeoCart'2008, held in Auckland, New Zealand on September 1st to 3rd, 2008. The conference is endorsed by the International Cartographic Association (ICA) and jointly organised by the New Zealand Cartographic Society, the University of Auckland, the University of Otago and Land Information New Zealand.

The contributions to this volume report on initiatives at the cutting edge of cartography and geovisualisation, covering relevant themes such as interactive atlases, 3D display in urban and marine applications, egocentric mapping, visualisation of uncertainty, geovisual exploration and current and future issues in web cartography. Each chapter has been reviewed by three international experts from the Paper Committee.

The first section of the book includes three chapters grouped under the **Representation** theme. Together they cover both static and dynamic representation, on primary and secondary elements of the map. Unusually we will start with a study that focuses not on the map itself but on the map legend. The paper by Juliane Cron and co-authors (including keynote speaker Lorenz Hurni) describes the development of a dynamic legend to ease the use of interactive atlases, (or Multimedia Atlas Information Systems - MAIS), an unexplored topic in this context. They illustrate a prototype Atlas of Switzerland that implements this technology, called "Smart Legend", which can manage the complex content of MAIS using a "Power Tool". And then from land to sea, Andrew Ternes and co-authors outline the user-centred development (an approach shared with Cron's work) of a 3D virtual reality chart for hydrographic surveying. Led by the mariners who would eventually use the chart, challenges such as enabling navigation in featureless maritime environments and the design of 3D symbols that were faithful to the extant 2D chart were addressed. Moving specifically to urban 3D, Markus Jobst and co-authors argues for non-photorealistic rendering (NPR) in virtual 3D city models to enhance their effectiveness and expressiveness, ultimately in a decision-making environment. Their novel 3D visualization designs are derived from adapting 2D graphical core variables into a 3D context.

The second section - **Egocentric Mapping** - contains two chapters that investigate the individual-focused point of view in geographical visualisations. The paper by Phil Bartie and co-authors furthers the urban 3D theme and explores the natural link between the egocentric and Location Based Services (LBS) delivered on mobile devices. They report on a project that investigates the derivation of landmark visibility metrics and maps that could help an LBS provide information to the user in context, with the most visible items being the most relevant. In terms of value for designing representations for the mobile context, Bartie's paper is of a kind with Jobst's NPR, especially useful in the limited display space involved. The second paper that shares an egocentric approach is by Grant Carroll and co-author, describing a dynamic and interactive map interface that uses hierarchically nested automated cartograms to emphasise areas of user interest or focus. This effectively uses local distortion to create a multi-scale representation, producing a fisheye-like display. User testing indicated that the cartogram map was more efficient than other ways of handling multiple scales, such as use of inset maps and zooming in.

Following this section are three papers that illustrate the exploration of tangible and intangible geographical phenomena by visual means, grouped under the broad theme of **Geovisual Exploration of Uncertainty and Terrain**. Starting with the intangible, revealing geospatial uncertainty has long been a topic of much research attention and two updates are presented here. Julian Kardos and co-authors give an account of the metaphorical basis by which a resolution-based uncertainty visualization technique, the *trustree*, communicates choropleth map uncertainty to the map viewer. How metaphors work is inherently a cognitive question, which links this chapter with the preceding section, since the egocentric point of view is essentially an intuitive one for all of us. A test revealed that less resolution leads to the perception of more uncertainty rather than the opposite, and is more usable. Moving from attribute uncertainty to spatial uncertainty, Tomaž Podobnikar uses Monte Carlo (MC) error modelling (as does Kardos) as the basis of estimating boundary uncertainty, which is subsequently visualised in novel ways. One such innovation is using superimposed and additive MC realizations of boundary as a texture feature. This paper is notable in that it tries to also depict uncertainty associated with the interior of a polygon through the displacement of a regular matrix of squares. The exploration of a tangible aspect of reality rounds off this section, with Igor Florinsky visually describing the terrain of our nearest neighbours in the Solar System via morphometric variables. His planet- or satellite-wide maps reveal hitherto unidentified patterns (helical structures) on the surfaces of these celestial bodies. Looking back over the section there is a progression in geometric terms from lines (both Kardos's and Podobnikar's boundaries) to areas (Podobnikar's polygon interiors) to surfaces (Florinsky's global exploration).

The last section in the book is on **Web Mapping** which, it can be argued, is how the majority of people worldwide with access to a computer now get their cartographic fix. The section starts off with a thorough investigation by Nigel Stanger of metrics associated with several web mapping delivery techniques at different scales of data load. This is an issue of great relevance for creators and viewers of maps on the Web as displays of large data volume start to be enabled. Stanger found that the image-based methods of delivery scaled better than those based on HTML. The Afterword is by William Cartwright and co-authors and is unique in this book as it describes future plans for developing an "Affective Atlas", a collaborative atlas that harnesses the innovations embodied in Web 2.0. This is an atlas in the spirit of Wikipedia, the online collaborative encyclopaedia, which elicits similar questions. An example issue that may arise and needs to be addressed is allowing non-experts (in the case of the affective atlas, non-cartographers) to create and edit content. The speculative nature of this paper makes it a natural choice for the Afterword. It also ensures that papers on interactive atlases, an important aspect of the book underemphasized by its running order, bookend this volume.

We would like to acknowledge the efforts of the authors who responded to our call for papers. Their cutting-edge research contributes to the advancement of cartography and GIScience and presents a well balanced international view on these subjects. Their determination in getting the chapters ready, despite the tight deadlines, is very much appreciated.

We wish to express our appreciation to the international experts who formed the Paper Committee. Many of them were asked to work on short deadlines, but their in-depth reviews contained valuable comments and exceptional suggestions for us but most notably for the authors.

Special thanks go to Professor Pip Forer for suggesting an inspiring (and intriguing) title for both the GeoCart'2008 Conference and this book and to GeoSmart Ltd from the Auckland's North Shore to be the Conference Gold Sponsor.

The book would never see the light of day without the assistance of the Springer-Verlag publishing team at Heidelberg. In particular we wish to express our gratitude to Agata Oelschläger who patiently guided us through the intricacies of the publishing process.

Finally, we would like to thank our families for their support, understanding and patience during the last few months of "I have to work on the book this weekend"...

To Kirsten, Alex and Danny …

Pragnę również podziękować Tobie Iwono, za Twoje kochające serce, troskę i cierpliwość, i Wam Natalio i Wando za Wasz uśmiech, pogodę ducha i zrozumienie.

Antoni Moore
Igor Drecki

July 2008

Paper Committee

Gennady Andrienko	Frauenhofer Institute, Germany
Mark Billinghurst	University of Canterbury, New Zealand
Manfred Buchroithner	TU Dresden, Germany
Barbara Buttenfield	University of Colorado, United States
William Cartwright	RMIT University, Australia
Keith Clarke	University of California, Santa Barbara, United States
Jeremy Crampton	Georgia State University, United States
Urska Demsar	National Centre for Geocomputation, Ireland
David DiBiase	Pennsylvania State University, United States
Martin Dodge	University of Manchester, United Kingdom
Danny Dorling	University of Sheffield, United Kingdom
Igor Drecki	University of Auckland, New Zealand
Jason Dykes	City University, United Kingdom
Robert Edsall	Arizona State University, United States
Sara Fabrikant	University of Zurich, Switzerland
Pip Forer	University of Auckland, New Zealand
Mark Gahegan	University of Auckland, New Zealand
Georg Gartner	Vienna University of Technology, Austria
Lorenz Hurni	ETH Zurich, Switzerland
Bin Jiang	University of Gavle, Sweden

For Bruce McLennan

CHAPTER 1

Facilitating the Handling of Interactive Atlases by Dynamic Grouping of Functions – The Example of "Smart Legend"

Juliane Cron, Samuel Wiesmann and Lorenz Hurni

Institute of Cartography, ETH Zurich, Zürich, Switzerland

Abstract

Over the past few years, interactive atlases have been extended by new functions and interactive techniques. Zooming and panning, switching layers on/off, changing classifications and colouring, as well as interactive queries are standard features nowadays. Related multimedia information such as diagrams, text, images, videos and animations, are linked to the geographic entities. Interactive atlases turn more and more into sophisticated Multimedia Atlas Information Systems (MAIS). At the same time, it has become increasingly difficult for atlas editors and developers to arrange this impressive wealth of information, functionalities and interactions in a user-friendly way. A possible means for enhancing the usability is the sophisticated grouping of functions in so-called "Power Tools" within the Graphical User Interface (GUI) of the atlas. As an example of such a tool, the transformation of the legend into a "Smart Legend" is shown. Unlike other atlas components, the functionality of the legend in interactive maps has hardly been touched by developments in the past few years, and the advantageous methods of interactivity are mostly omitted in map legends. The great hidden potential of such a tool can crucially influence the development of new atlases in the future.

Keywords: interactive atlas, legend structure

1 From Digital Maps to Interactive Multimedia Atlas Information Systems (MAIS)

Apart from the digital production of conventional maps, the introduction of Geographical Information Systems (GIS) and other computer aided cartographic production systems in the 1980s has also enabled the development of the first fully computer screen-based map products. Throughout the years, the richness of the geographical data that can be included, the degree of interactivity, analysis and visualisation capabilities has meant that the overall quality of the graphical design of both the maps and the Graphical User Interface (GUI) has significantly increased. Unlike most GIS, modern electronic maps and atlases are based on cartographically edited and symbolised data (instead of raw GIS data) and specifically selected, implemented, and designed functions and interactions. In addition to map data, thematic data and related multimedia data can also be included. According to Hurni (2008) such Multimedia Atlas Information Systems (MAIS) are systematic, targeted collections of spatially related knowledge in electronic form, allowing user-oriented communication for information and decision-making purposes. Typical applications of MAIS are Reference (World) Atlases, School Atlases, National and Regional Atlases, Topographic Atlases, Thematic Atlases, and Statistical Atlases (Hurni 2008). In the last few years the main focus of most of the products was centred on granting access to more data and functions, bringing MAIS close to "regular" GIS which unfortunately still have a limited usability. The guidance of the user through the almost unlimited number of possible functions had rather low priority. Most functions were mainly arranged in more or less structured menu lists. In the following, a new approach is presented, which structures and groups the functions in a logical, thematically-oriented way, but also allows the adaptive combination of interplaying functions.

2 The Basic Framework: Classification of MAIS Functionality

Besides maps and data, MAIS also offer special atlas functions, allowing geographic and thematic navigation, querying, analysis and visualisation both in 2D and 3D mode. These interactive functions, in the following referred to as atlas or map tools, enable the user to manipulate and visualise maps and data. Thus, generally speaking, their functionality enables interactions with the geographic and thematic content of the MAIS.

Driven by the rapidly changing needs of the "information society", the number of interactive atlases is steadily growing and they are getting more powerful in terms of thematic content, functionality and interactivity. Due to the fact that these atlases are developed for various target groups, they have different scopes, quality levels and data content. Accordingly, the applications offer a wide variety of functions, e.g. navigation functions are important for world atlases, and data analysis and visualization functions are crucial parts of statistical atlases. Table 1.1 shows the most important functions, arranged in five main groups (general functions, navigation functions, didactic functions, cartographic and visualisation functions as well as GIS functions).

In the past, rather divergent approaches for the classification of functions in interactive atlases were presented (e.g. by Ormeling (1997) and Bär and Sieber (1997)). This is an indication of the difficulty of setting up a comprehensive, well-founded grouping or classification of the functions. Table 1.1, showing possible functions in interactive atlases, is the result of an analysis of the most popular and comprehensive atlas products currently available. The order and integration of the functions in those atlases only partly correspond to the other classifications mentioned above. The existing compilations were therefore modified and extended by newly found functions. Particularly, new function subgroups consisting of individual functions were introduced (e.g. the subgroup "redlining"); and some functions were reassigned to other groups or subgroups. The new classification was set up to serve as a basic framework for the proposed development of a user-friendly atlas GUI by means of functional segmentation.

Table 1.1. Classification of functions in MAIS (Cron et al. 2007)

Function Groups	Function Subgroups	Functions
General functions		Mode selection, language selection, file import/export, printing, forward/ backward, placing bookmarks, hot spots, settings (preferences), tooltips, help, display of system state, imprint, home, exit
Navigation functions	Spatial navigation	Spatial unit selection, move map (pan, scroll), enlarge/reduce of map extend (zoom in/out, magnifier), reference map/globe, map rotation, line of sight and angle, determination of location (coordinates, altitude), placement of pins, spatial/geographical index, spatial/geographical search, tracking
	Thematic navigation	Theme selection and change, index of themes, search by theme, favourites
	Temporal navigation	Time selection (selection of time period, positioning of time line), animation (start/stop etc.)
Didactic functions	Explanatory functions	Guided tours, preview, explanatory texts, graphics, images, sounds, films
	Self-control functions	Quizzes, games
Cartographic and visualisation functions	Map manipulation	Switch on/off layers, switch on/off legend categories, modification of symbolisation, change of projection
	Redlining	Addition of user defined map elements, addition of labels (labelling)
	Explorative data analysis	Modification of classification, modification of appearance/state (brightness, position of sun), map comparison, selection of data
GIS functions	Space and object-oriented query functions	Spatial query/position query (coordinates query, query of altitude), query of distance and area/ measurement, creating profile
	Thematic query functions	Thematic queries (data/attribute queries), access to statistical table data
	Analysis functions	Buffering, intersection, aggregation, terrain analysis (slope, exposition, etc.)

3 The Implication of MAIS Functionality on the GUI Design

In order to meet the demands of a broad user community, the potential of MAIS should not be restricted only to various multimedia visualisation techniques. Moreover, the programmes should also encourage the user by means of simple and intuitive handling, which is mainly enabled by a well-structured arrangement of the different map tools on the computer screen. The presented classification of the map tools therefore also aims for a corresponding segmentation of the GUI. This means that functions of the same group or subgroup should be arranged in close proximity to each other.

It is primarily the atlas GUI which allows the user to activate the functions. The various functions are integrated in the atlas GUI by controls like buttons, lists, sliders, etc. As one of the main parts of an interactive atlas, the GUI exists in order to bring all components (map, data and functions) together. Commands are usually not initiated by keyboard entry or natural speech, but by the selection and movement of the elements on the Graphical User Interface (direct manipulation) (Shneiderman and Plaisant 2005).

The main purpose of a GUI is to supply users with a perfectly tailored set of tools, enabling them to interact in a comprehensible and undisturbed way with the system. While communicating using this visual desktop environment, the interactive application should advise or even guide the users to the relevant information. In other words, the information and the specific tools should be presented to the users by means of a well-structured GUI. The success of an application depends to a great extent on its usefulness and usability. Thus, a GUI should be laid out in such a way that the intuitive and associative structure of human thinking is supported. According to Marinilli (2002), the design of a GUI is therefore conducted using seven basic principles of visual communication:

- Know your user
- Minimise the short term memory load on users
- Preserve consistencies
- Ensure overall flexibility, error recovery and customisation
- Follow standards
- Make explicit the system's internal state
- Provide the user with the sense of control

A GUI should therefore follow a user-centred "thin design" approach containing organisational, economic and communicational aspects (Marcus et al. 1994): According to the *organisation principle*, the conceptual structure of the GUI should be simple, clear and consistent. The *economy principle* states that there must be a maximisation of the user's and the system's efficiency with a minimal set of content-related tools. The *communication principle* asks for a customisation of the presentation with respect to the user's intake capacity. As a further component, we would like to add the *aesthetic principle*, where graphical issues, such as colour schemes and the level of iconisation of all GUI elements are of great importance (Cron et al. 2007).

MAIS usually consist of extensive geographical, thematic and statistical information. The situation can be compared to a general store containing thousands of goods. From the vendor's point of view, the question is how to sell these articles best with a maximum presentation effect and minimal costs. From the buyer's point of view, the question is rather how to access all the desired articles in the shortest time. Using this analogy, one of the main questions in atlas GUI authoring is therefore how to structure and design a GUI to attract the user's attention and to transmit the relevant information to the user. Today, as the number of atlas applications increases, there is also a great variety of GUI solutions.

Once again, the foremost aim of atlas GUI design is a comprehensible presentation of the atlas screen. To obtain a well-balanced composition of the screen area, the segmentation into functional-logical units – analogous to the function groups – is applied (Fig. 1.1). Functional units help to minimise the graphical "load" of the GUI and facilitate the user's orientation in order to interpret and operate the whole set of atlas functions in an easier and faster way.

Since an atlas is understood as a formal compilation of maps, the map should always dominate the overall layout and the GUI respectively. This requires an appropriate segmentation of the screen area, a strict partitioning of the GUI elements, ingenious interaction possibilities between GUI elements, and a high consistency of content, graphics, actions and feedback (Sieber and Huber 2007). Segmentation is accomplished in a functional sense. According to the basic framework, Figure 1.1 shows the implementation of the atlas function groups by corresponding atlas segments. For example, the spatial navigation segment includes a reference map, coordinate display, etc.

This approach of GUI segmentation according to the atlas function groups seems reasonable with respect to the use of the atlas. Thereby, the usability of an application will surely be increased. However, due to the static nature of this approach, it is limited when applied to a MAIS which is extensive both in content and in function. Here, the static approach cannot meet all demands for implementing the function integration.

Fig. 1.1. Atlas GUI, segmented in functional-logical units (Prototype “Atlas of Switzerland”, Version 3); 1. General functions, 2. Navigation functions, 3. Didactic functions, 4. Cartographic and visualisation functions, 5. GIS functions

In most cases, the function groups and subgroups do not include purely self-contained functions. Therefore, an alternative approach is important as well: the interplay of functions (Sieber and Huber 2007). Functions may technically belong together, although they are not assigned to the same function group with respect to their functionality. For example, the use of one tool entails the involvement of further tools belonging to other function groups or subgroups (e.g. in the 3D-part of the “Atlas of Switzerland”, the manipulation of the tool “line of sight and angle” requires the use of the “redraw-button” which is part of the “preview” tool).

Furthermore, the classification of the functions is based on the fact that the user carries out a manipulation in the atlas. Wiesmann (2007) calls these functions “user-oriented functions”. However, in this context an additional consideration must be the reaction of the system: Some functions are executed by the system without being noticed by the user; another question is which functions are provided by the system at all due to various constraints. Wiesmann (2007) calls these functions “system-oriented functions”. In addition, there are also functions which are mutually exclusive due to the nature of the basis data (Schneider 2001). To ensure both the user-friendliness and an extensive functionality of the atlas, the GUI must become more flexible, dynamic, and self-adapting in such cases. This means that due to their assigned function group or subgroup, the atlas tools cannot be integrated in the GUI in the same strictly segment-based way.

4 Adaptive Grouping of Functions for the “Smart Legend” Concept

By applying the concept of the dynamic combination of functions mentioned before, this chapter illustrates the advantages of this flexible approach using the example of the legend. This adaptive arrangement combines several user- and system-oriented functions from different function groups and subgroups which interplay appropriately. Individual functions can also be compiled in a hybrid way: depending on the situation, the function is directly applied by the user or is triggered internally by the system in order to support the user. The examples in this chapter illustrate how such a situational interplay of functions could work.

The GUI acts as an interface between the user and the MAIS. Similarly, the map legend supports the communication process between the cartographic sign language, which encodes the spatial information of the map, and the user, who wants to read the information. The legend is an important part of the map. Its main task is to explain the individual map elements supporting the user’s perception of the message of the map. It is therefore the key to the map content. Without the legend, the map information cannot properly be understood and in some cases is entirely unintelligible.

In order to allow the legend to function as an interface, the map symbols must be unambiguously designed and clearly distinguishable from each other. It is obvious that the explanatory symbols in the legend must match the symbols in the map in order to guarantee a clear assignment. However, it turns out that these requirements are even more difficult to fulfil on a computer screen than on paper. Many digital legends do not satisfy these criteria.

Also other authors considered the digital legend in their research. In the case of animated maps, Kraak et al. (1997) as well as Peterson (1999) presented interactions within the legend to support the user. Miller (1999) suggests “self-describing symbols” which provide their own descriptive information, Lobben and Patton (2003) organise the layer structure hierarchically and provide additional information about each layer when accessed, and van den Worm (2001) presents a "pop-up web map legend" – which is a realisation of Miller's (1999) suggestion mentioned above – as well as a "control-panel web map legend" which allows to switch on/off layers. All the mentioned approaches deliver important inputs for legend design, but concentrate on particular situations where the suggested solution is adequate. None of them focused on the behavioural pattern of the legend in an overall concept. Bollmann and Koch (2002) go one step further: they state various functions and point out the possibility of the interactive legend to act as the central access point to the map data, to other media, as well as reacting in its appearance on the user's actions. But they also narrow the concept down to some isolated theoretical examples.

The functionality of the legend, which will be demonstrated in the following, was developed within the framework of a "user centred development" in the sense of a "rapid (paper) prototyping" (Preim 1999, Snyder 2003) with the support of 40 subjects. In two iteration loops the subjects had to rate on a 5-step scale from "not useful at all" (-2) to "very useful" (+2) a total of 60 different kinds of behavioural and functional patterns of the legend. These ratings were statistically analysed (e.g. Wilcoxon signed-rank test) with a level of significance of 5%. The detailed test descriptions and results were reported by Wiesmann (2007).

The goal was to achieve a more efficient legend, here called "Smart Legend", taking the user's needs better into account. The term and the idea behind "Smart Legend" were introduced by Sieber et al. (2005). In this paper, the conceptual ideas are revised, expanded, and transferred into the development process described above.

The "Smart Legend" distinguishes itself by a variety of interactions, which encourage the user to engage with the map's content. These interactions should simplify the interpretation of the map information. Both the handling of the legend and the implemented interactions should be user-friendly and intuitively designed. The single interactions are not isolated from each other but they build a unity. This allows the user to step from one option to the next without even noticing. The ingenious combination of functions within the "Smart Legend" supports and accompanies the user while working with the map. In the following, the main characteristics of the "Smart Legend" will be clarified.

The first short example recalls the approach of GUI segmentation according to function groups from the previous section. Displaying and hiding a map layer several times in a row ("switch on/off layers") may facilitate the reading and comprehension of the map information. Therefore, it may be helpful to integrate the "switch on/off layers"-function directly into the legend. However, in many electronic atlases this user-oriented function is only available on a separate panel. As a result, to toggle between thematic layers, the user has to switch between the legend and the panel.

However, the mentioned carried out tests indicate that the subjects would rather like to have this function directly available within the legend (Fig. 1.2). Bollmann and Koch (2002) suggest this function should be executed by clicking on the legend label or box. On the example in Figure 1.2 a separate checkbox is used because the legend label and legend box themselves consist of further integrated functions. In case of an ambiguity when perceiving the assigned symbols or colours, the displaying and hiding of a layer may be of use by means of simplifying the user's decoding process.

In order to improve the main task of the legend, a combination with another function might be considered: The "spatial query" command is a function, which could be implemented explicitly in the user-oriented mode, but it can also run in the system-oriented mode, i.e. without direct user interaction, as in our case. The function can be described as follows: After each zooming and panning the system queries the current map extent and verifies which map objects are displayed within that area. The "Smart Legend"-function processes the result of the query and reacts accordingly.

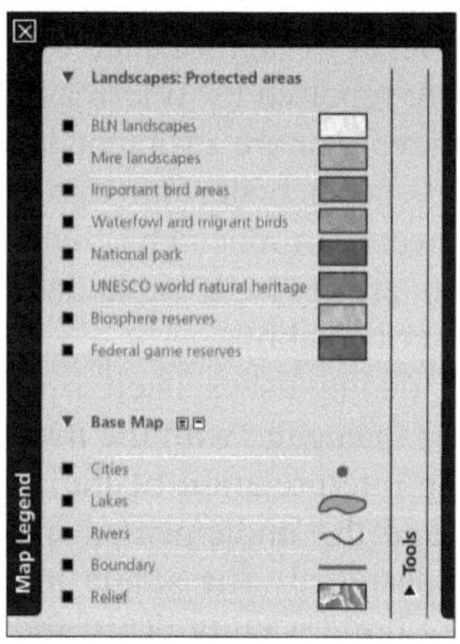

Fig. 1.2. Functions and regular legend elements combined within the legend: checkbox to toggle the map layer, legend label, and legend box (explanatory symbol)

The purpose of this procedure is based on the fact that in a MAIS in particular, complex maps with numerous layers – and therefore also a large number of symbols – are often used. However, when navigating through the map, and especially when working in detailed zoom levels, only a few of these symbols remain in the actual map window. In a traditional map legend this fact is not considered. The mentioned tests showed that there is a need for action. Interestingly, the subjects clearly distinguished between elements of the base map and elements of the thematic map layers:

On one hand, the elements of the base map usually serve as orientation aids for the map user. According to the results of the tests, these elements should only be listed in the legend, if they appear on the current map extract. Base map elements, which are outside of that window, should not be listed and therefore not explained in the legend (see e.g. Figure 1.3: the layers "cities" and "lakes" disappeared). By clicking an extra button (to the right of the title "base map"), the legend is expanded and the non-selected base map layers are available again.

On the other hand, elements belonging to one of the thematic map layers were evaluated differently. Here, all elements should be listed in the legend, even if they were not currently visible. However, the elements outside of the window and therefore not visible on the map, should be dimmed in such a way that the user would be able to recognise at a glance, which of the thematic layers could be expected in the map window (Fig. 1.3). The layer "national park" for example is not within the map window and therefore its label is dimmed. "BLN landscapes", on the other hand, is within the actual map extent and thus it is not dimmed. Thematic layers, which are intentionally hidden (=switched off) by the user, should not only be dimmed in the legend (legend label) but the explanatory symbol should not be displayed at all (see e.g. layer "mire landscapes" in Figure 1.3). By doing so, the number of explanatory colours is reduced and wrong allocations may be avoided.

Furthermore, the functionality of the legend can be expanded by implementing another useful function: "bidirectional highlighting" is based on a concept which was already proposed by Huber and Schmid (2003). On one hand, the corresponding legend entry of an object of the map is automatically highlighted on mouse-over (Figure 1.4a: an object of "important bird areas"). On the other hand, all objects in the map belonging to the same object class are emphasised when a legend field is selected. This can be achieved in two ways: by highlighting all map objects belonging to the selected layer (Figure 1.4b: layer "waterfowl and migrant birds") or by dimming all non-selected layers.

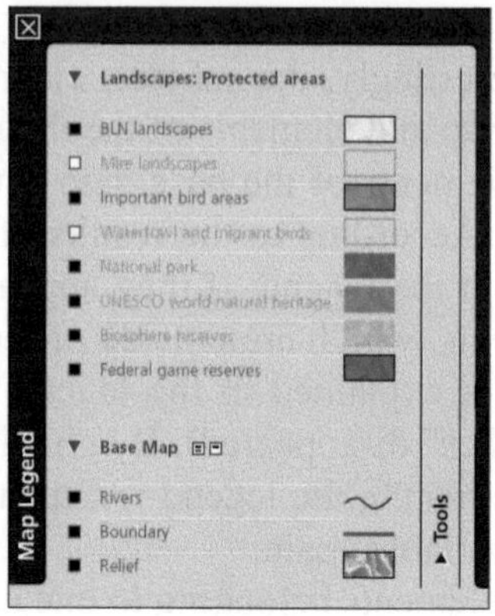

Fig. 1.3. Appearance of the "Smart Legend" as a result of querying the current map extent

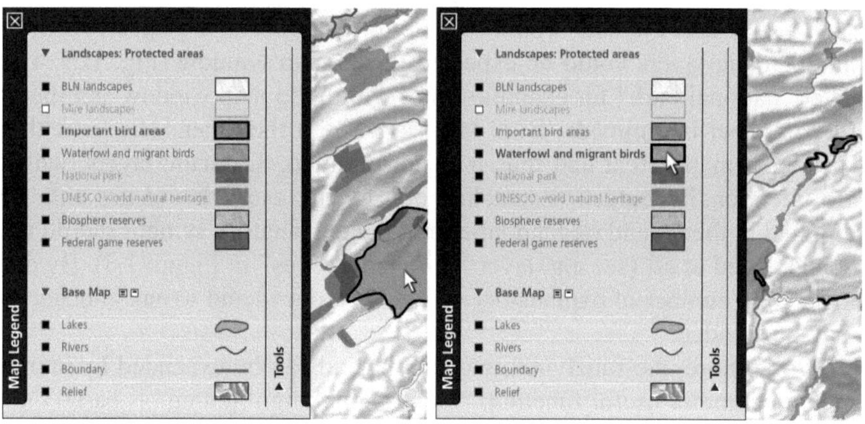

Fig. 1.4a and b. "Bidirectional highlighting" on the map and in the legend

Hence, one legend entry may perform three functions simultaneously, i.e.:

1. "switch on/off layers" and "switch on/off legend categories" respectively (by means of a checkbox);
2. dimming of all non-selected layers in the map (by clicking the label);
3. highlighting of all map objects of the selected layer (by clicking the legend box).

Functions 2 and 3 represent two alternative ways of implementing the function "selection of data". The latter (3.) is especially suitable for layers containing rather small objects: without highlighting these objects could barely be detected on the map.

The usefulness of the legend may be further increased by carefully designing the individual legend entries as follows: Apart from using exactly the same colours, also the opacity of the elements should correspond both in the map and in the legend. In the case of coloured map areas with a shaded relief in the background, the depiction of a background relief is recommended in the respective legend box. In case of geometric map symbols, their size should be adapted in the legend in such a way that it corresponds to the symbol size appearing on the map, even in various zoom levels.

All the functions described above can be implemented in an interactive, compact legend which hardly requires more space than a conventional, static legend. The subjects considered it very important that the legend is kept as small, simple and well arranged as possible. The map image should be emphasised and the subjects therefore preferred versions where the legend could be hidden in favour of increased space for the actual map. Furthermore, they considered it sensible to adapt the size of the legend to its content. The legend window will automatically get smaller when it only contains few entries and will enlarge if necessary. However, the assignment of a stationary location and a constant size – which can lead to almost empty or overfilled legends depending on the map topic – was not considered helpful.

Despite this request for a small and compact legend, further user-oriented functions in the legend turned out to be highly desirable according to the test results. Such a legend with extended functionality would not harm the advantages of the "Smart Legend" concept; however, they should be made available as options. This extended legend could be accessible step-by-step, by "walking through" a compact, hierarchically structured version of the legend; in other words: the "Smart Legend" can be expanded on demand. In this way, the legend turns into a veritable "Legend Centre" of the MAIS. The figures above show a legend with an extension button "Tools" by which this "Legend Centre" is accessible: the power-tool "Smart Legend" is available (Fig. 1.5). The above mentioned smart functions remain available and unchanged.

In such a "Smart Legend Power Tool" the functions mentioned above can be combined with further user-oriented functions. The functions belonging to the groups "map manipulation" and "explorative data analysis" were rated to be very useful and desirable by the subjects; in particular the modification of colours. Other candidates for combination can be found in the groups "GIS-functions" (e.g. "terrain analysis" or "creating profile") or "didactic functions" (e.g. "graphics" or "explanatory texts"). Figure 1.5 for example, shows a possible extension of the functions enabling the choice of colour, shape, size and arrows, as well as an optional extension for displaying statistical information ("diagrams").

It would also make sense to further develop the idea of user- and system-oriented functions in this extended "Smart Legend Power Tool" environment. In this way, as long as no layer is selected, all the available functions are dimmed – which means inactive. As soon as the user selects one layer, the system checks which one has been selected and it automatically offers all applicable functions, depending on the selection. At the same time, the current graphical attributes of the layer objects are exported and set accordingly within the adaptation tools of the legend. In the test, such extended system-oriented functions were considered to be very helpful by the subjects. In Figure 1.5 for example, the layer "federal game reserves" is selected, a layer consisting of closed areas. Accordingly, the functions "choice of colour", "adaptation of transparency" and "diagrams" are available; the current colours and transparency are sorted from the map. The options "changing the shape", "changing the size" or "adding arrows" remain inactive. In the case of the selection of a layer with point objects, the options for the choice of shape and size would be active as well.

The settings selected by the user are immediately applied to the map by the system (without clicking a redraw button). At the same time these settings are transferred to the legend in order that the information in the map and the explanation always coincide. This is applicable for all adaptations. Thereby the points and lines are not only enlarged in the map, or changed in shape, but also adjusted in the legend. On the example in Figure 1.6 the layer "rivers" is selected, therefore the function "changing the size" is available (additionally to the "choice of colour" and "adaptation of transparency"), and the stroke-width of the rivers is enlarged in the map and the legend simultaneously. This is the only way to make sure that the legend always provides the optimal explanations of the map.

As one can see, functions of different subgroups can be combined in a user- or system-oriented mode without harming the usability of the MAIS; sometimes just the opposite is found to be the case: The combination of different functions in a "Power-tool" turns out to be highly desired by the users, it is technically feasible and it supports an intuitive handling of the atlas. However, despite all such smart functions, the emphasis when implementing a MAIS should still be laid on a proper map design and not on sophisticated technical features.

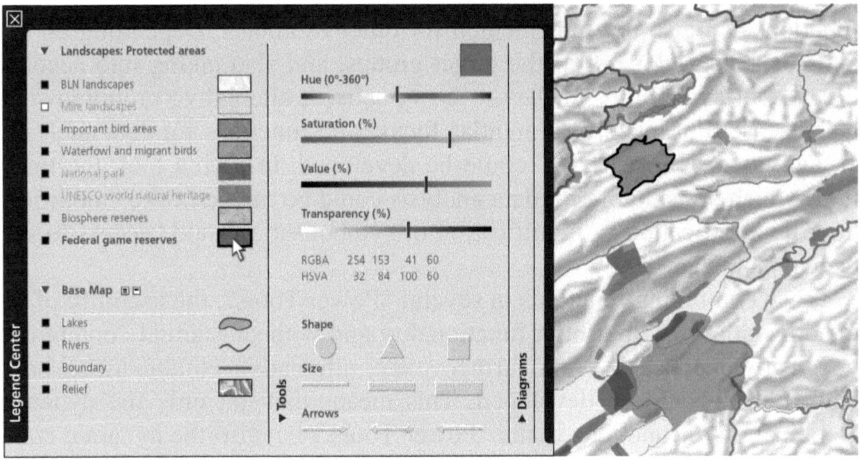

Fig. 1.5. “Smart Legend Power Tool”: The system decides which functions are active and which are not

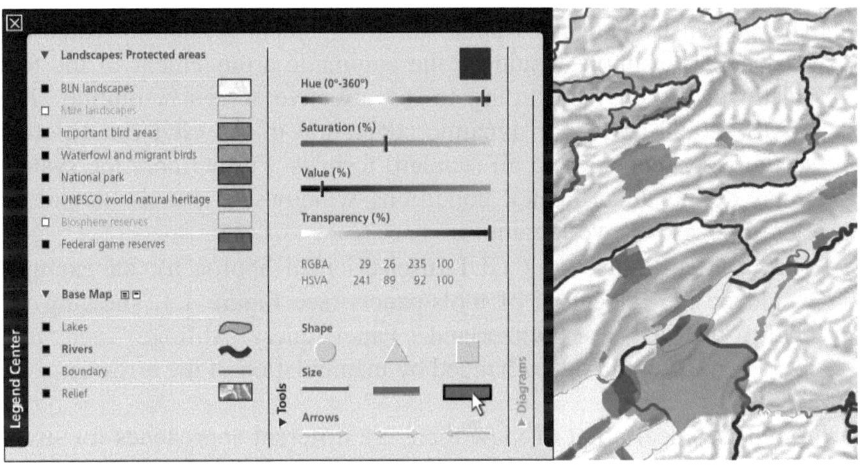

Fig. 1.6. Adapted offer of applicable functions. Changed settings are immediately applied to the map and to the legend

5 Conclusions

The example of the “Smart Legend” shows that a more flexible, dynamic and self-adapting combination of functions, coming from different function groups and subgroups, facilitates the use of an interactive atlas and therefore increases its usability. However, individual combinations of functions can also be implemented for other examples. Depending on the aims of the developers, of the target groups, and also taking into account the current availability of data to be visualised, alternative groupings can make sense. For instance if an atlas focuses on functions for data analysis, an individual "Power Tool" could be developed. In such a tool, the functions which are suitable for data analysis could be merged (like "modification of classification", "modification of symbolisation" and "access to statistical table data").

When combining functions in several "Power Tools", the usability of interactive atlases with a high functionality and with substantial content can be greatly increased. As a further stage, alternative combinations on a higher level could be developed. This means that not only the dynamic combination of functions in the "Power Tools", but also the dynamic combination or flexible arrangement of the "Power Tools" themselves within the entire atlas GUI is possible. Thereby, again two different ways of flexible arrangement can be distinguished:

- The system itself decides upon the arrangement (system-oriented approach); an example would be the automatic arrangement of the tool palette in the GUI according to the effective screen resolution. The MAIS should allow for dynamic adaptation of the GUI for different screen sizes, considering all standard formats. Furthermore the system could define under which conditions windows or panels should be closed in case of overlapping.
- The user customises the GUI (user-oriented approach); an example would be the relocation of tools/panels (see Figure 1.1, showing the GUI of the "Atlas of Switzerland": Panel "query (Abfrage)" and "topics (Thema)" could be exchanged by manipulation of the arrows).

From a technical point of view there are different approaches for structuring the functions in a MAIS. Apart from the technical feasibility, a major difficulty for the atlas developer lies in choosing a suitable combination of user- and system-oriented functions. He also has to determine under what circumstances a specific function – e.g. for changing settings or for modifications – is requested by the user, and in the framework of which tool this function can be arranged in the most usable and intuitive manner. It is also recommended that the potential users of the atlas should be included in the development at an early stage, in order to better assess their specific needs.

References

Bär HR, Sieber R (1997) Atlas of Switzerland – Multimedia Version: Concepts, Functionality and Interactive Techniques. Proceedings of the 18th International Cartographic Conference, Stockholm, Sweden, pp 1141-1149

Bollmann J, Koch WG (eds) (2002) Lexikon der Kartographie und Geomatik, vol 2. Spektrum Akademischer Verlag, Heidelberg

Cron J, Sieber R, Hurni L (2007) Guidelines to Optimized Graphical User Interfaces of Interactive Atlases. Proceedings of the 23rd International Cartographic Conference, Moscow, Russia

Huber S, Schmid C (2003) 2nd Atlas of Switzerland: Interactive concepts, functionality and techniques. Proceedings of the 21st International Cartographic Conference, Durban, South Africa

Hurni L (2008) Multimedia Atlas Information Systems. In: Shekar S, Xiong H (eds) Encyclopedia of GIS. Springer, Berlin Heidelberg New York, pp 759–763

Kraak M-J, Edsall R, MacEachren AM (1997) Cartographic animation and legends for temporal maps: Exploration and or interaction. Proceedings of the 18th International Cartographic Conference, Stockholm, Sweden, pp 253-260

Lobben AK, Patton DK (2003) Design guidelines for digital atlases. Cartographic Perspectives, 44:51-62

Marcus A, Smilonich N, Thompson L (1994) The Cross-GUI Handbook: For Multiplatform User Interface Design. Addison Wesley, Boston

Marinilli M (2002) The Theory behind User Interface Design, part 1, http://www.developer.com/design/article.php/1545991

Miller S (1999) Design of multimedia mapping products. In: Cartwright W, Peterson M, Gartner G (eds) Multimedia Cartography, 1st edn. Springer, Berlin Heidelberg New York, pp 51-63

Ormeling F (1997) Functionality of Electronic School Atlases. Proc of the Seminar on Electronic Atlases II, ICA Commission on National and Regional Atlases

Peterson MP (1999) Active legends for interactive cartographic animation. International Journal of Geographical Information Science 13(4):375-383

Preim B (1999) Entwicklung interaktiver Systeme: Grundlagen, Fallbeispiele und innovative Anwendungsfelder. Springer, Berlin Heidelberg New York

Schneider B (2001) Integration of Analytical GIS-functions in Multimedia Atlas Information Systems. Proceedings of the 20th International Cartographic Conference, Beijing, China

Shneiderman B, Plaisant C (2005) Designing the User Interface: Strategies for Effective Human Computer Interaction. Addison Wesley, Boston

Sieber R, Huber S (2007) Atlas of Switzerland 2 – A highly interactive thematic national atlas. In: Cartwright W, Peterson M, Gartner G (eds) Multimedia Cartography, 2nd edn. Springer, Berlin Heidelberg New York, pp 161–182

Sieber R, Schmid C, Wiesmann S (2005) Smart Legend – Smart Atlas! Proceedings of the 22nd International Cartographic Conference, A Coruña, Spain

Snyder C (2003) Paper prototyping: the fast and easy way to design and refine user interfaces. Morgan Kaufmann Series in Interactive Technologies, San Francisco

Wiesmann S (2007) Smart Legend – Aufbau und Interaktivität digitaler Kartenlegenden. Diploma thesis, University of Zurich, Department of Geography. http://www.ika.ethz.ch/teaching/Diplomarbeit-Wiesmann.pdf

Worm J van den (2001) Web map design in practice. In: Kraak M-J, Brown A (eds) Web Cartography: Developments and prospects. Taylor and Francis, London

CHAPTER 2

A User-defined Virtual Reality Chart for Track Control Navigation and Hydrographic Data Acquisition

Andrew Ternes[1], Peter Knight[2], Antoni Moore[3] and Holger Regenbrecht[4]

[1]Hydrographic Survey & Spatial Data, Port of Melbourne Corporation, Melbourne, Victoria, Australia

[2]School of Surveying, University of Otago, Dunedin, New Zealand.

[3]School of Surveying / Department of Information Science, University of Otago, Dunedin, New Zealand.

[4]Department of Information Science, University of Otago, Dunedin, New Zealand.

Abstract

In hydrographic surveying dynamic electronic displays are used to assist in navigating the survey vessel to a desired waypoint or maintain a desired track. The visual display for the coxswain (the "driver" of the ship) is typically a two dimensional (2D) planimetric view which is oriented so that the vessel direction is facing up on the display. This can be confusing at times for the coxswain, as he / she has to relate the virtual desired track on the display to the surrounding spatial reality, sometimes with few references to aid the process. The coxswain has to mentally visualise the desired track as an invisible line across the sea surface and attempt to steer the vessel down the line. The objective of this project is to investigate the use of a three-dimensional (3D) nautical chart with intended users of the chart involved as a focus group at the design and feedback stages of its development. The intent of the project is to create a virtual environment that will aid in track control and other forms of navigation. It is proposed that the chart will increase the spatial and situational awareness of the mariner and allow intuitive and quicker decision making in running desired tracks for hydrographic surveys. The project makes use of a perspective view from the helm of a virtual vessel, modelled on the actual vessel used to trial the system. The mariners in the user focus group agreed that the 3D chart was effective, creating an enhanced cognitive appreciation of where the vessel was in relationship to its surroundings, while faithfulness to the 2D chart was also commented on.

Keywords: three dimensional nautical charts, virtual marine environment, user focus groups, hydrographic survey, seamless bathymetry-topography, Electronic Nautical Charts (ENC), navigation

1 Introduction

Hydrographic survey operations routinely utilise electronic charts to visually display desired tracks allowing a regular spacing of sounding lines to cover the seafloor. The importance of maintaining track during survey operations is critical to upholding hydrographic standards. Therefore the coxswain has to comprehend accurately and precisely what he/she perceives on a navigation display in order to make a decision on how much steerage needs to be applied to maintain the desired track. Furthermore, a coxswain must be able to locate beginnings and ends of survey lines and be able to switch between runlines and runline blocks in a timely manner.

Currently in hydrographic surveying the positioning of vessels is run in a two dimensional (2D) space. A global positioning system (GPS) is coupled with a computer running an electronic chart. This enables the vessel's position to be displayed and constantly monitored by the master alongside other instrumentation to maintain position or track. Orienting and positioning a vessel being displayed in 2D, while remaining aware of the three dimensional (3D) reality surrounding the vessel, can sometimes be confusing to the coxswain. The stimulus for developing this technology comes from the experience of observing coxswains attempting to get on and to maintain track during hydrographic surveys in areas where there might be few references connecting reality with the 2D display.

The main objective of the research is to demonstrate the practical use of a virtual reality display of a 3D nautical chart, in which a perspective view from the helm of a virtual vessel can be used. In theory, creating and implementing a 3D nautical chart should increase the spatial and situational awareness of the coxswain, and allow intuitive and quicker decision making in running desired tracks for hydrographic surveys. The chart will have a 3D baseline dataset of sonar-derived singlebeam and multibeam bathymetric data. It will be enhanced by survey-accurate virtual points of reference (e.g., navigational aids) to enhance the associations between the desired track and reality. Also, shore topography and some land-based vector features will be included in the virtual model.

With any form of land based virtual environments for 3D maps, users usually have multiple points of reference, such as buildings and landmarks for self-orientation and positioning. The situation is similar at sea, as many mariners refer to shore-based points of reference, such as lighthouses and lead beacons. Additionally, when far offshore or in times of poor visibility (such as at night or in fog), certainty in position is reduced, leaving the mariner to rely upon GPS, radar and compass. In such circumstances, track control may be extremely difficult. This is where a virtual beacon or lead system within a virtual environment with a helm perspective, might be of benefit. These visualisations together with other virtual navigational information could form the basis of a 3D chart that a vessel could use for navigation. Although such charts have been implemented before (e.g. Ford 2002), the development of the chart at the centre of this project is user-led. Mariners were consulted at the initial design stage and the prototype development stage (which bookends the stages of development reported on in this paper).

The next section of the paper will be a short review of hydrographic charting with specific reference to 3D charting. This will be followed by a fuller presentation of the project scope and aims. The methods section will then cover the consultation with the user focus group, development of the virtual environment and dataset; and linkage with realtime GPS. Results and feedback from a focus group will be reported on before discussion and conclusions.

2 Review

2.1 2D Charts

The first electronic 2D nautical chart was NAVSHOALS, produced in 1976 from digitised charts (Ward et al. 2000; Ford 2002). Since then, Electronic Nautical Charts (ENCs) have become an almost universal navigational tool, perhaps as significant to the history of navigation technology as the implementation of radio[1]. The International Maritime Organisation (IMO) and the International Hydrographic Office (IHO) together set international standards for the production and display of Electronic Chart Display and Information Systems (ECDIS). Many users however, rely on ENCs produced by private sector suppliers, in formats which, while not conforming to the IMO-IHO standards, are nevertheless often supported by the national hydrographic agencies because of their manifest usefulness (Pasquay 1996).

2.2 The 3rd Dimension and 3D Charts

The benefits of 3D display have long been acknowledged and are intuitive. Support for use of the third dimension is offered by Swanson (1999, p 183), who noted that 3D display engages "...a larger portion of the brain in the problem solving process." This support for 3D display echoes with the findings of Porathe (2006), who saw 2D displays as being less efficient than a 3D display in test conditions. If there is an improvement overall, this would contribute to an enhanced situational awareness on the part of the user (mariner) in relation to the tasks that they have to do (hydrographic operations and general nautical navigation), "... [awareness] of what is happening around you and understanding what information means to you now and in the future" (Endsley et al. 2003, p 13). Techniques for establishing usability of the display (e.g. Nielsen 1993; Hackos and Redish 1998) can indicate whether such awareness is being fostered.

However, the inability of the 3D view to show all the viewable information at one time has to be taken into account. The supporting use of a 2D representation has been suggested for assessing layout and distance estimation, two tasks that are difficult to do in 3D (Ware and Plumlee 2005).

The initial idea of three dimensional digital nautical charts was conceptualised in 1976 at the time of the development of NAVSHOALS (Ward

---

[1] Captain Pace of Canada Steamship Lines addressing the Canadian Hydrographic Service in 1995

et al. 2000; Ford 2002). Ford reasoned that 3D charting would improve safety and efficiency of transportation on water. Gold et al. (2004) added that such a realistic simulation "...should be an advantage in navigation and training". Ford's (2002) 3D chart allowed the collection, processing and application of nautical data, incorporating the use of bathymetric, topographic (height – TIN data), environmental and nautical data (e.g. real-time tide and current data) in the formation of an integrated three dimensional visual system (Ford 2002). Although supported by differential GPS to give abilities approaching 4D (i.e. real-time update), the chart and topographic map are raster scans of paper-based media.

Gold et al. (2004) developed a 3D "Pilot Book", providing navigational assistance for ships (particularly non-local ships) entering Hong Kong harbour. They developed an application that utilised interactivity and dynamics (of viewpoint, objects and relationships – including collision detection) attributes, already well-developed in games development. Froese and Wittkuhn (2005) reported on an application that was conceptualised as a 3D chart (a European project known as the Electronic Pilot Display and Information System - EPDIS). Responding to a call by mariners for new navigation technology to take advantage of recent innovations, a project was undertaken to provide a 3D navigation product aimed at serving pilots – in this case pilots operating in the Baltic and Mediterranean Seas.

The use of 3D and stereoscopically enhanced 3D visualisations is explored and summarised in the hydrographic context by Ostnes et al. (2004a, 2004b). Their papers present evidence from fields such as aviation and geophysical science to support the superiority of 3D and enhanced 3D visualisation over planar views for hydrographic applications.

2.3 Virtual Reality

"A virtual reality is defined as a real or simulated environment in which a perceiver experiences telepresence" (Steuer 1992, p 76) or in more technological terms as a computer-generated, three-dimensional, interactive system. Virtual reality (VR) was and is successfully applied in different domains such as automotive design review, military training or geological surveying. There is strong reason to believe, that (tele-) presence in such a virtual environment leads to good orientation, wayfinding and awareness. Thus defined, it can be regarded as the interface through which interaction with 3D models such as those outlined in the last section, takes place.

VR can either be of the desktop variety or immersive (Fisher and Unwin 2002). With desktop VR, the virtual environment is delivered through a conventional computer set-up and in an immersive environment the user experiences the virtual world with specialised equipment such as head-mounted displays or large, stereoscopic, multi-display setups. Even though the immersive method would seem the most realistic VR experience, it needs expensive equipment and can induce disorientation. Most of the time, the same information can be delivered effectively using desktop VR, which induces none of these effects (Fisher and Unwin 2002). For the mariner's scenario, immersion would also represent a step away from the real world, where constant vigilance is needed for navigation. Context may also be important for navigating through virtual environments (see Purves et al. 2002, for an exploration of this theme in mountain environments), implying that virtual content for an experienced mariner may differ in certain aspects from virtual content for a recreational boat user (though both examples have to adhere to maritime regulations).

Bodum (2005) has categorised virtual environments according to their degree of realism and time dependency. Taking a marine navigation beacon as an example, the former criterion can vary along a continuum from a realistic depiction of that specific beacon, through indexed representation (where the beacon, depicted through a generic example of its type, is embedded in a semantic structure), rendering as icons (a simplified beacon retaining pictorial meaning – this is the 3D type used in this example – see 4.3.1), symbolisation (a more abstract explicitly cartographic representation) to language-based depictions (description of the beacon, for example as set out in the NZ Pilot).

Time dependency can be of static (specific space at a specific time), dynamic (model is fixed in space but information within changes) and real-time types ("…coordination of changes in a model that can affect the representation in several different spaces at exactly the same time", Bodum 2005). According to this definition, the application reported on in this paper lies somewhere between dynamic (e.g. the action of a flight simulator) and real-time (update from differential GPS, or dGPS, operating in real-time though there are no other autonomous objects in this virtual environment).

2.4 The Use of Focus Groups

The assessment of virtual applications, being an example of an interface, is normally accomplished through usability testing (e.g. Nielsen 1993; Hackos and Redish 1998; Endsley et al. 2003). It is user-focused and often the cognitive affinity of the user with the interface is being assessed. User focus groups have been used for such evaluations and are a cost-effective way of generating qualitative data (also as a preliminary to a more intensive usability assessment – Harrower et al. 2000). Monmonier and Gluck (1994) sees them as "…useful in addressing a broad range of cartographic problems".

In the cartographic or visualisation domain they have often been used to assess interfaces, with members of the group first exploring the interface to assess effectiveness, then expressing their feedback to the developer (Monmonier and Gluck 1994; Harrower et al. 2000; Lucieer and Kraak 2004). We have used several mariners from Dunedin and Melbourne to form our user focus group. Although there will only be a limited review by way of analysis, such an approach can be effective (Monmonier and Gluck 1994).

3 Aim and Scope

3.1 Aim

The aim of the research is to create a 3D model of the real world environment to which virtual navigation aids may be added. This model can then be displayed to represent the coxswain's field of vision as if seen in a forward looking camera image, similar to the perspective of the coxswain's own field of vision in reality. This would inform the mariner of the vessel's location and orientation. The master of the vessel will then have an immediate reinforced perception of his/her position in reality and in relation to the chart. The technique utilises more recognisable associations between 3D chart objects and real world objects thereby closing the conceptual gap between the real world's environmental surrounds and thc electronic chart.

Currently trying to maintain track with 2D displays requires the master to consult the chart, check the compass, check the surrounding environments, pick a point of reference to steer towards, then correct the helm (Figure 2.1, left flow chart). Creating a 3D chart display in combination with vessel heading and 3D objects, such as virtual leads, the master will not have to consult the chart in order to apply steerage, all he/she needs to do is steer towards the point of interest on the virtual display (Figure 2.1, right flow chart). This eliminates two steps in the reaction process, as three processes are combined into one.

It is envisaged that the result of the project can be applied to many aspects of hydrography from the positioning of oil rigs, pinpointing geological sample sites, and aiding navigation; to general shipping, navigating dangerous waters, and collision avoidance in periods of bad visibility such as bad weather, fog and darkness.

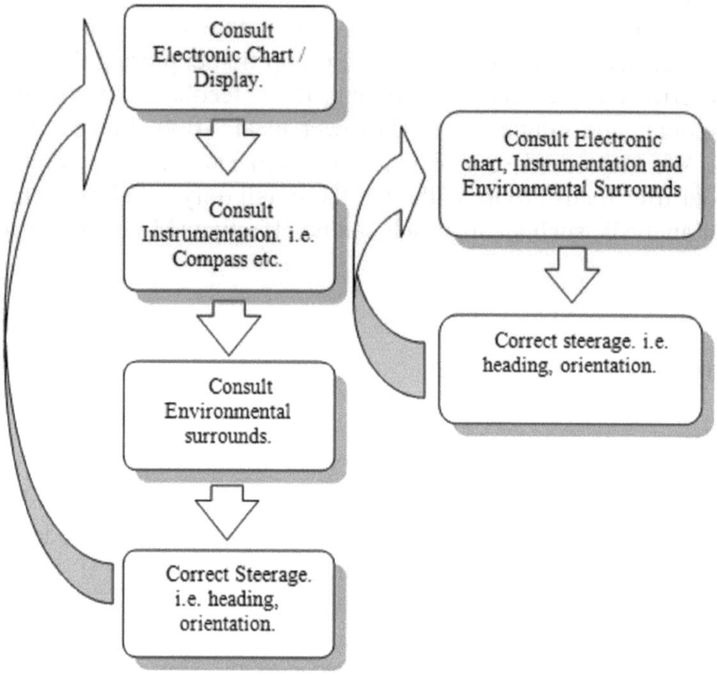

Fig. 2.1. The process on the left follows the method used to apply steerage using a 2D display. The proposed system (right), utilising a 3D chart, will attempt to reduce this process

3.2 Project Scope

In order to gather information relevant to the design of a 3D chart, a questionnaire was given to vessel handlers with experience in hydrographic surveying. These experienced mariners formed the focus group that would inform the design of the 3D application. Insight was sought into the importance to the helmsman of each piece of navigation equipment and also how well that equipment currently performs. Also identified were current problems in displaying navigation information, and ideas were considered concerning how this information might be presented more effectively.

The 3D model which was eventually developed combined information from the nautical chart (including bathymetry) and, unusually for a hydrographic application, elevation information of the surrounding land area (topography) to create a digital elevation model (DEM). All of the relevant navigation objects (marks, buoys, shoreline features) were digitised off a 2D nautical chart and transformed into similar 3D symbols in the virtual space, which were designed with International Hydrographic Organisation (IHO) 2D symbol standards in mind. The area modelled is Otago Harbour near Dunedin, New Zealand including its surrounds up to the harbour entrance.

In order to use this model for navigation a perspective was created which represents the real world vessel's perspective. This was achieved by creating a digital (virtual) model of one such vessel from vessel diagrams. The virtual vessel is geographically referenced dynamically, by regular updates in real time through GPS fixes on board the vessel. The master then has a visualisation of the 3D chart in real time from a similar perspective to that of a helmsman's view in the real world.

By creating a 3D chart, populated by virtual reference beacons (working in the same way as lead lights or range lights) as well as the vessel itself, it is intended that the navigation of the vessel to a desired location or maintaining the vessel on track will be greatly facilitated. The focus group was returned to after implementation of the demonstration 3D chart, to collect feedback and recommendations (by discussion / interview) for the next stage of chart development

4 Methodology

4.1 User Survey Results

The requirements and specifications are the base upon which the model is initially designed. It is therefore important to gain preliminary user information. For this project surveys and questionnaires were conducted with four masters and skippers of Otago University School of Surveying (OUSS) and the Port of Melbourne Corporation (PoMC), forming an experienced user group (with most of the participants having at least 15 years survey experience). Only the OUSS staff were involved with the implementation, interpretation and testing of the application. Summarising their nature of experience, all members of the focus group possessed domain (marine or hydrographic) expertise, not VR expertise.

The questions that were chosen for the survey were formulated to assess; what is liked and disliked about current electronic navigation charts; how effective current electronic navigation charts are; what information is critical to maintain track; which instruments mariners rely on the most; what information they would like to see added to an electronic navigation chart; how effective mariners think a 3D chart would be; and finally how electronic navigation charts and navigation information might be improved.

The results of the questionnaire show that mariners were pleased with ENCs and find them useful for making immediate decisions. However, the mariners cautioned that ENCs must be set up properly and / or verified with a paper chart or RADAR. The mariners showed great interest in the possibility of viewing the ENC in three dimensions. Questionnaire results indicate a 3D model must comply with the following:

- Be useful for making immediate decisions
- Aid the mariner with his / her visual and spatial awareness.
- Be in a defined projection working in conjunction with the GPS.
- Be able to plot vessel position and plot vessel tracks and waypoints.
- Be able to display vessel heading, course made good, speed over ground, compass bearings and depth possible from the echo sounder.
- Have enough information for safe navigation.
- Have clear symbology representative of the current standard symbols and good representation.
- Visual display of vessel in relation to dangers, obstructions and land.
- Must not overwhelm the mariner and distract from the navigation from the vessel.

These principles guided the implementation and were revisited with the user group after the demonstration system has been developed. An appendix containing the questions posed and summarised feedback is provided after this paper.

4.2 Datasets

The gathering of reference material came from a multitude of sources. For example maps, topographic and bathymetric data sets, building and vehicle designs, photos, video, recordings, sound and light sources. For this project topographic data, vessel data, beacon data, reference photos and nautical charts were used for the creation of the virtual environment.

The structuring of the virtual model was done in the 3D visualisation software application Fledermaus™ (Interactive Visualisation Systems 3D – IVS 3D 2008). The data for the surrounding topography of Dunedin is a 20m Digital Elevation Model (DEM) which was acquired by point elevations and digitisation of 20m contours from the Land Information New Zealand (LINZ) topographic chart series. This Universal Transverse

Mercator (UTM) coordinate system was chosen for this and all subsequent datasets, predominantly because the metre unit would facilitate 3D object dimensioning but also for hardware and software compatibility. Data for the project were obtained from surveys by the Port of Otago and by digitising depths from the current nautical chart of Otago Harbour (NZ 6612). Additional data for certain areas around the harbour surveyed over the years (by OUSS) were also added to the combined dataset. All datasets were converted from their existing projections, coordinate systems and datums to WGS84 UTM (Zone 59 South). This is a solution for visualisation purposes only, as additional complex rectifications to account for tide, for example, would have to be made to create a dataset that can support navigation. The conversion and compilation was accomplished using ESRI ArcGIS and Fledermaus, with the resulting DEM having a final resolution of 10m.

Colours for the 3D chart were copied from NZ 6612 with the bathymetry closely resembling the colours and contours associated with that chart. (see Figure 2.2 and Figure 2.3). This was to maximise affinity of the 3D chart with the mariners who predominantly use the source 2D chart in navigation and associated and established conventions of colour and symbology.

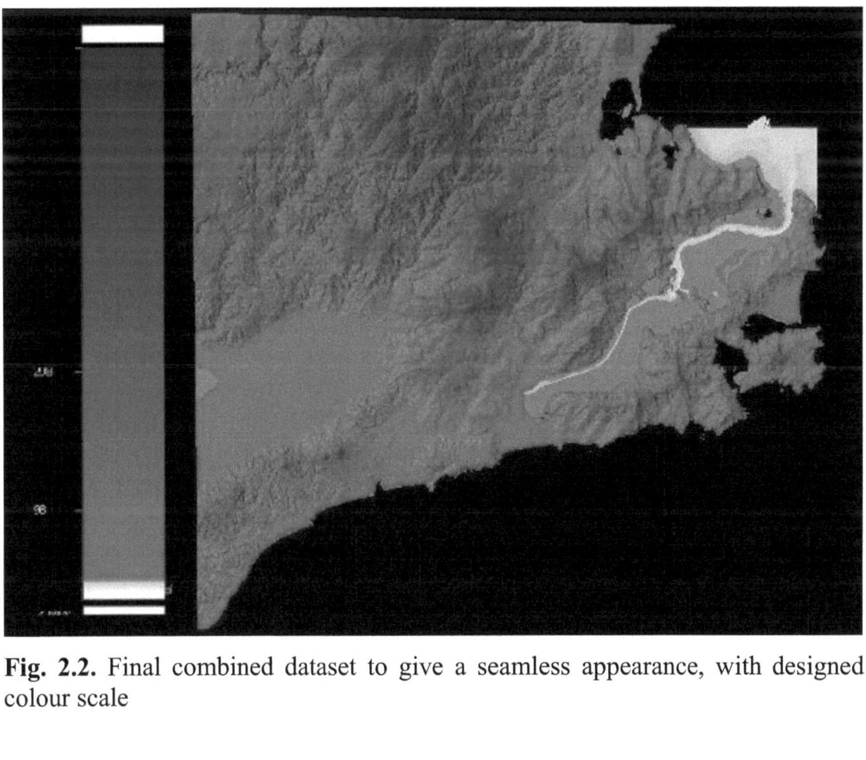

Fig. 2.2. Final combined dataset to give a seamless appearance, with designed colour scale

Fig. 2.3. Perspective view seen from the top of Otago Harbour. The city of Dunedin surrounds the head of the navigable harbour channel (in light shade)

4.3 3D Objects

Construction of objects from the reference material was performed in Autodesk 3D Studio Max 9™, which is a three dimensional model builder which can be coordinate-based, and so helped facilitate the positioning of objects within Fledermaus™. The objects constructed fall into two categories, the 3D navigation aids (such as beacons and lead lights) and the virtual vessel itself.

4.3.1 3D Navigation Aids

Often in track control there are no navigation aids present. To address this, virtual reference beacons or leads were created for the master to steer towards (i.e. a virtual lead light or range light). This system uses two lights or signals, one located above and one a distance further behind the first lead. When the leads align, the boat is in the navigation channel (Toghill 2003).

Additional key navigation information needed to be identified and designed. This included the design of beacons, light houses, leads, and other principal marks. The symbols on chart NZ 6612 were standardised using the International Association of Lighthouse Authorities (IALA) Maritime Buoyage system – Region A (Red to port) (see Figure 2.4). Using NZ 201 symbols terms and abbreviations (HO of RNZN 1995), the chart NZ 6612 and photos of the actual beacons, three dimensional renderings (which would be regarded as "iconic" on Bodum's realism-abstraction scale, 2005) were created using basic primitive shapes within 3D Studio Max 9™ (e.g. see Figure 2.4a, b and c). Gold et al. (2004) drew from IHO standard Electronic Navigation Charts for the design of their symbols for navigational features such as bouys, lights and underground rocks. These were positioned using their known coordinates and were designed to mirror reality as closely as possible, but with appropriate abstractions (Figure 2.5 and 2.6).

Fig. 2.4a. IALA Port Beacon, **b.** Real Port Beacon, **c.** 3D Port Beacon

Fig. 2.5. A view from the vessel in reality

Fig. 2.6. A view from the virtual vessel within the model

a

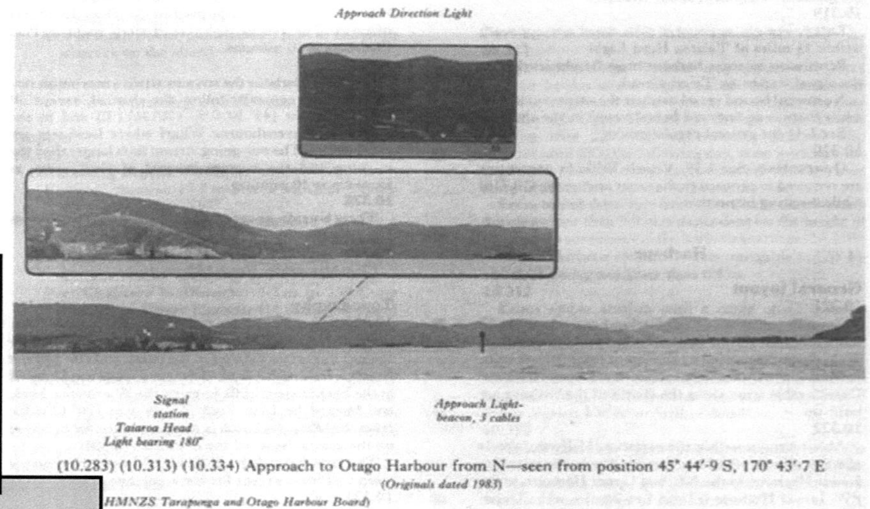

b

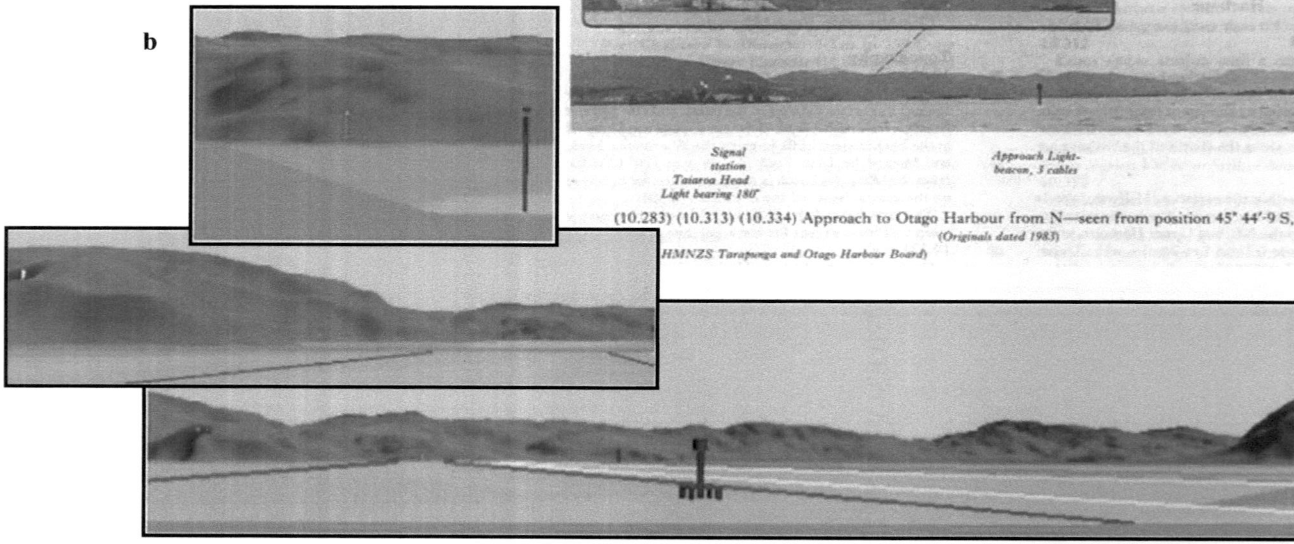

Fig. 2.7a. The approaches as displayed in the New Zealand pilot (Watts 1987). Taiaroa Head and the Approaches light beacon are signified, **b.** The simulated view of the New Zealand Pilot approaches

A key aid during this process was the chart NZ 6612 and the New Zealand pilot guide. Using the New Zealand Pilot as a guide, an attempt was made to model the approaches and coastline features as closely as possible. This meant including as many of the principal navigation marks that are used around Otago harbour as possible. Among those included were the church steeple at Port Chalmers, the memorial on Memorial Hill, the lighthouses at Taiaroa Head and Heyward Point, the East Cardinal beacon on the Mole, and the Approaches light beacon (see Figure 2.7). These landmarks and other aids to navigation were represented by abstracted 3D symbols. The attempt to simulate the NZ Pilot is another strategy to strengthen the mariner's affinity with the 3D representation, which as the review has shown, has been known to lead to incomplete data display. Gold et al.'s (2004) 3D marine GIS implementation draws from the Hong Kong pilot book as source, the challenge being to render as 3D a "graphics-free description of a simulated Real World" (p 21).

4.3.2 3D Vessel Design

A three dimensional model of the University of Otago's vessel, the *MV Polaris* was also designed in 3D Studio Max, using refurbishment plans, reconnaissance photos, and hand measurements of deck features. This virtual vessel was imported into Fledermaus. The helm of this 3D model Polaris provided the dynamic real time viewing point with the 3D chart (see Figure 2.8), enabled by GPS.

Fig. 2.8. The real Polaris (left) and virtually (right)

4.4 GPS Integration

The virtual vessel's position was obtained dynamically in real time from MV Polaris' differential GPS. The differential GPS Antenna was mounted on the Polaris approximately amidships on the roof of the cabin. The location of the antenna was measured with an offset tape to get an approximation of its position. These values were required to ascertain the position of the camera view from helm of the 3D model Polaris. Fledermaus has a function in its vessel manager function where it can input a National Marine Electronics Association (NMEA) string that carries a GPS data message. This allows for the positioning of the virtual vessel to match the real vessel in real time. A view can then be generated from the helm of the virtual vessel. The master then has a visualisation of the 3D chart in real time from a similar perspective to that of the helm in the real world.

4.5 Post-implementation Focus Group Interviews

A discussion and interview with the mariners following the field test (see next section for details of this) took place to gain feedback on the performance of the 3D chart. In a question-answer format the mariners were asked how well the 3D chart displayed information, and whether the information was clear and easy to interpret. Like the initial questionnaire, the feedback was recorded on paper. This feedback formed their qualitative evaluation of the 3D chart in relation to the real-world environment it was trying to depict.

5 Results and Discussion

A final demonstration of the system was organised, with the original Otago focus group (section 4.1). The objective of the demonstration was to set up a survey run line similar to that of a normal Hydrographic survey and use the model and the perspective from the virtual helm to navigate (see Figures 2.9, 2.10 and 2.11). The set up of the survey line to be run included the use of some virtual beacons along the survey line transect. These were set up to aid the helmsman maintain track.

Many mariners when running survey lines judge how far they are off the desired track by using a port and starboard offset indicator. What is lacking, however, is a point of reference to steer towards. A mariner will typically look at the bearing of the desired track and try to steer that bearing on the compass while looking towards a distant object on land. When at sea and far from land, the mariner has no point of reference to orientate themselves. It was hoped that by placing a virtual beacon within the scene the mariner will obtain a point of reference to spatially orientate themselves.

Fig. 2.9. The real-world setting of the 3D chart (on the laptop). The model is used to navigate the vessel along a survey line

The test went well, with the master steering using the model to navigate down the survey line (see Figure 2.12). However, there was some latency (time delay) in the system, which caused the virtual vessel to be slightly hard to steer.

Latency and slightly awkward steering were due to a number of factors, such as the GPS heading being based upon the previous recorded positions, rather than an accurate observed heading (this could be improved by using a dual GPS antenna arrangement for accurate heading observation). The model was also run on laptop which is not as well-suited as most desktops for running high power graphics. There were also hardware issues (i.e. GPS-laptop data connection) contributing to the latency. Despite these limitations the system ran well using what was available and showed that the concept can work.

Feedback from mariners was obtained by an interview and discussion process. A replay of the field test was displayed to two members of the focus group, emphasising the other capabilities of the software, in particular, the features which can aid navigation when run in conjunction with a 3D nautical chart. The mariners agreed that the display was extremely effective in conveying the location of the vessel with respect to real and virtual objects. The mariners both agreed that the display helped to create a better and greater understanding of location and attitude of the vessel, particularly when it was necessary to navigate to a geographic position by coordinates without a real world point of reference.

Fig. 2.10. Too far to port, so the virtual beacons are not in line so the vessel needs to steer starboard. View is from the helm of the 3D model of Polaris

Fig. 2.11. The virtual beacons are in line, so the vessel is on the desired track. View is from the helm of the 3D model of Polaris

The mariners both liked the presentation of the model and thought that it related approximately to that of the official paper chart NZ 6612. The task of creating the model presented a challenge, particularly in relation to the representation of the symbology for the navigation aids. The beacons had to closely resemble reality so the cognitive association can be made between the real beacons and the virtual beacons. However, the 3D beacon symbology also had to be conveyed in such a way that the beacon would be clearly identified and associated with the real world beacon as well as with the symbol on the chart NZ 6612. The real and symbolised objects were quite different, making it difficult to design a 3D symbol that resembled both. The decision was made to closely copy reality as a starting point, so all the associations between the real world and the model can be made by the mariner (in this way the mariners' feedback aligned with the aims of the research group in developing a 3D application). Using the chart NZ 6612 all necessary navigation information was then incorporated into the model, such as, leading lights, navigation markers and principal marks. The International Hydrographic Organisation (IHO) is already setting standards for electronic navigation chart symbology, and it is hoped that in time, these could be extended to 3D symbology standards as well (Froese and Wittkuhn 2005).

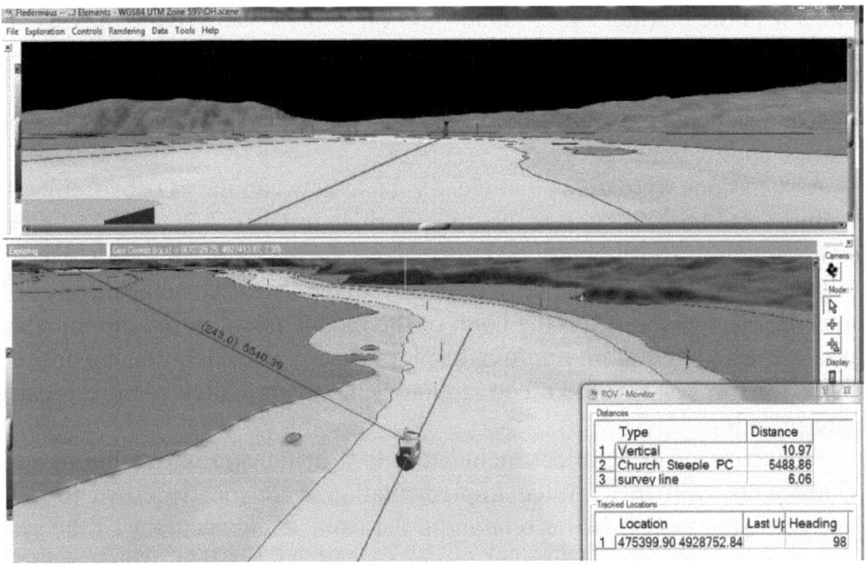

Fig. 2.12. The display used to navigate the vessel. The bottom display shows a remote 3D perspective view, while the top view shows the model from a user defined perspective from the helm of the 3D vessel model

6 Conclusions

The user-led implementation of a 3D nautical chart for any given area can be advantageous to mariners, hydrographers, and geologists, to name but a few. The literature reviewed for this research project clearly showed the potential effectiveness of the use of a three dimensional chart used in real time. The literature also provided a guideline or general procedure to follow in the construction of this kind of virtual chart. The project constructed a virtual chart of Otago Harbour using 3D software, employing numerous sources for datasets, including digitised chart information, digitally acquired data, and unusually for such an application, Digital Elevation

Models (DEMs) of land topography (included in deference to the role of land-based landmarks for navigation in coastal waters). The 3D or virtual chart was constructed and designed with mariner feedback and suggestions leading the development. The information was obtained by questioning mariners who are familiar with existing methods of navigation for hydrographic surveys. The information provided by the mariners provided the insight to what to include and display within the virtual chart.

Operational demonstrations of the use of the virtual chart in hydrography and education of GIS proved to be successful. Interviews and discussions with the mariners after demonstration of the system found the display to be an improvement upon existing implementations. The mariners found the 3D chart to be advantageous in making decisions for hydrographic survey and general navigation and great potential was foreseen for diverse marine applications.

Associations were quick and clearly made between the 3D chart objects, features and hazards with reality. Similarities to Chart NZ 6612 and clear representation were made and aided the formulation of one's spatial awareness without distraction. For example, the use of virtual beacons and a bridge perspective from the helm of the model vessel in the virtual chart was identified as a more intuitive display as one can clearly see the desired track for the vessel, unlike conventional 2D charts which require greater interpretation.

The scope of this project included only a demonstration of the model. Constraints with this limited implementation were mostly down to data and hardware issues. Issues relating to the latency and accuracy of the system were mainly due to limited resources and information. There was no pre-existing reference frame for the Polaris, as it was recently acquired by the university. The design of the vessel model was based upon hand measurements and old refurbishment plans. This allowed the basic model to be constructed and used for demonstration purposes.

Differential rather than real time kinematic (RTK) positioning was used. This provided a good solution for horizontal position but not vertical. Future research could implement a RTK system to obtain real time tide and study the depth underneath the keel, making for a truly dynamic (4D) virtual model and refining the accuracy of the seamless bathymetry-topography surface. The GPS used for the survey also only had a heading based on the velocity of the GPS antenna. The heading could be greatly improved by using a dual antenna system GPS compass (an echo sounder would also be useful).

It was noted with the construction of the virtual chart that the New Zealand pilot guide was very useful in establishing the navigation aids, principal marks and sailing directions to be incorporated into the virtual chart. It was also noted that for future development, a survey or questionnaire could be circulated amongst a larger sample of the local marine operators on Otago Harbour. This would help identify the unofficial, but useful, principal marks, such as buildings and other features on the skyline that are commonly used. These could then be modelled and positioned in the virtual chart.

Consultation with such participants may also include usability testing (Endsley et al. 2003) for navigational tasks to simulate operative and quantitative assessment of the system. One such test would have the mariner verbalising their thoughts and actions as they perform the same navigation task action with both the 3D virtual system and the traditional system. This is collection of qualitative usability data by think-aloud protocol (Hackos and Redish 1998), which can subsequently be analysed by techniques such as emergent themes analysis (Wong and Blandford 2002). This would move the emphasis to an assessment of the simulation itself, rather than satisfying the requirements of field testing that we have seen here.

The application of augmented reality is also suggested for displaying a three dimensional nautical chart in the field of view of the mariner. The possible addition of real time images from a video camera was also theorised to be a useful application to hydrography. However, given the accuracy issues with the 3D boat model, a proper vessel reference frame still has to be constructed before these innovations can happen.

The benefits from a virtual chart (and possible implementation with augmented reality applications) are significant. Spatial awareness of obstructions, geographical locations, desired tracks and routes in reference to the vessel are greatly increased. This may facilitate intuitive and quicker decision making, which is highly important in positioning a vessel or equipment in hydrographic surveying, offshore mining, fishing and scientific research.

Acknowledgments

First, the authors would like to thank the members of the focus groups from the Port of Melbourne Corporation and the MV Polaris (Otago) for their invaluable feedback. We are also grateful to Hamish Horvath, Allan Sutherland and Mike Denham for the provision of data used in this research. Thanks to Captain Graham Turner, Alistair Neaves, Steve Taylor, Hugh Parker, Kim Arnup and Paul Downie for their help on hardware; Douglas Bergersen, Mark Paton, Brian Grant and Peter Benfell should be thanked for their software support. Finally, we are grateful to the reviewers for their useful recommendations for this paper.

If you would like a copy of the Fledermaus™ 3D .sd or .scene files for this application, please contact the lead author.

References

Bodum L (2005) Modelling Virtual Environments for Geovisualization: A Focus on Representation. In: Dykes J, MacEachren AM, Kraak M-J (eds) Exploring Geovisualisation. Elsevier, London, pp 389-402

Endsley MR, Bolte B, Jones DG (2003) Designing for Situation Awareness. Taylor & Francis, London

Fisher P, Unwin D (2002) Virtual Reality in Geography: An introduction. In: Fisher P, Unwin D (eds) Virtual Reality in Geography. Taylor & Francis, London, pp 1-4

Ford SF (2002) The First Three-Dimensional Nautical Chart. In: Wright D (ed) Undersea with GIS. ESRI Press, Redlands, CA, pp 117-138.

Froese J, Wittkuhn D (2005) EPDIS – Electronic Pilot Display and Information System. The Hydrographic Journal 116:3-5

Hydrographic Office of the Royal New Zealand Navy (HO of RNZN) (1995) NZ 201 (INT 1) Symbols, Abbreviations, Terms used on Charts, 3rd edn. s.l.: Hydrographic Office of the Royal New Zealand Navy

Gold C, Chau M, Dzieszko M, Goralski R (2004) 3D Geographic Visualization: The Marine GIS. In: Fisher PF (ed) Developments in Spatial Data Handling. Springer, Berlin, pp 17-28

Hackos JT, Redish JC (1998) User and task analysis for interface design. John Wiley, New York

Harrower M, MacEachren A, Griffin A (2000) Developing a Geographic Visualization Tool to Support Earth Science Learning. Cartography and Geographic Information Science 27(4):279-293

Interactive Visualization Systems (IVS 3D) (2008) Images. Available from http://www.ivs3d.com/images/images2.html (accessed: 30th May 2008)

Lucieer A, Kraak M-J (2004) Interactive and visual fuzzy classification of remotely sensed imagery for exploration of uncertainty, International Journal of Geographical Information Science 18(5):491-512

Monmonier M, Gluck M (1994) Focus Groups for Design Improvement in Dynamic Cartography. Cartography and Geographic Information Systems 21(1):37-47

Nielsen J (1993) Usability Engineering. AP Professional, Morristown, NJ

Ostnes R, Abbott V, Lavender S (2004a) Visualization Techniques, An Overview – Part 1. The Hydrographic Journal 113:3-7

Ostnes R, Abbott V, Lavender S (2004b) Visualization Techniques, An Overview – Part 2. The Hydrographic Journal 114:4-9

Pasquay JN (1996) The Electronic Chart and Its Provisional Substitutes. International Hydrographic Review LXXIII(2):17-25

Porathe T (2006) Improved situation awareness in navigation using egocentric view 3-D nautical charts. In: Pikkar RN, Kooningsveld EAP, Settels PJM (eds) Proceedings of the 16th International Ergonomic Association's World Congress, Elsevier, Holland, p 6

Purves R, Dowers S, Mackaness W (2002) Providing context in virtual reality: the example of a CAL package for mountain navigation. In: Fisher P, Unwin D (eds) Virtual Reality in Geography, Taylor & Francis, London, pp 175-189

Steuer J (1992) Defining virtual reality: dimensions determining telepresence. Journal of Communication 42(4):73- 93

Swanson J (1999) The Cartographic Possibilites of VRML. In: Cartwright W, Peterson MP, Gartner G (eds) Multimedia Cartography, Springer, Berlin, pp 181-194

Toghill J (2003) Traditional and Electronic Navigation: An Introduction to Methods and Equipment. New Holland Publishers Ltd, Sydney, pp 49-51

Ward R, Roberts C, Furness R (2000) Electronic chart display and information system (ECDIS): State-of-the-art in nautical charting. Iin: Wright DJ, Bartlett DJ (eds) Marine and Coastal Geographical Information Systems. Taylor & Francis, London, pp 149-162

Ware C, Plumlee M (2005) 3D Geovisualisation and the Structure of Visual Space. In: Dykes J, MacEachren AM, Kraak M-J (eds) Exploring Geovisualisation. Elsevier, London, pp 567-576

Watts DT (1987) New Zealand Pilot, 14th edn. Hydrographer of the Navy, Taunton, pp 383-392

Wong BLW, Blandford A (2002) Analysing ambulance dispatcher decision making: Trialing emergent themes analysis. In: HF2002, Human Factors Conference "Design for the whole person – integrating physical, cognitive and social aspects", Melbourne, Australia

Appendix: Focus Group Questions and Summarised Feedback

Initial Questions and Summary of Answers

Question 1: What do you like most about the way navigational information (on electronic and paper charts) is displayed?

The general response to this question was that the data displayed on Electronic Navigation Charts (ENCs) is good for quick and immediate reference to the mariner, often in association with the RADAR. It was commented that although paper charts are very accurate and more reliable, ENCs are more immediate, but are often verified with RADAR and paper charts. One mariner commented that there was less of a need to refer to the paper charts when using an ENC particularly for plotting vessel position and laying off courses. In reference to navigation for surveying, the coast file is often lacking information for safe navigation. Paper charts and some ENCs use standardised symbols and presentation, which allow for clear interpretation, this was a particular representation that was also commented on.

Question 2: What do you like least about the way navigational information (on electronic and paper charts) is displayed?

When using paper charts alone there is an amount of time needed to plot the position on the chart, and there is always the possibility of an error occurring, whereas ENCs are more immediate. For any given chart the soundings are always obsolete and some paper charts have not been updated from the original survey. This can sometimes be transferred to the ENC if it is based upon a paper chart. Negative aspects of ENCs are that they are hard to keep updated. For example notice to mariners' corrections are difficult to update on an ENC. There is also a lack of information, such as buoys, and beacons around some secondary ports and harbours which makes it difficult for some mariners. There also appears to be differences between standard symbols and representation with varying ENCs and this is a concern for the mariner, as it can often make it confusing when trying to verify the information on a paper chart. There seems to be some lack of faith in the production and use of ENCs. There have been instances in marine reports where vessels have accrued damage due to improper use and set up of their ENC (Graham Turner, pers.comm). Some of the main areas of danger in the set up of an ENC relate to use of projections and matching these with the incoming GPS information. Incorrect projection or GPS datum can cause shifts in the position and misinform the mariner of the boat's actual real world position.

Question 3: What do you rely on the most to maintain position and track as regards navigation displays, for example heading, speed over ground, course made good?

The navigation displays that mariners mostly rely on to maintain track are heading; course made good; Radar (at night); GPS chart plots; paper chart plots; their own local knowledge of the work area, and the use of a ENC. Current ENCs being used by the mariners questioned are MaxSea and Cmap.

Question 4: What do you rely on the most to maintain position and track as regards the instrumentation, for example compass, radar, etc.?

The instrumentation that mariners mostly rely on to maintain track are the echo sounder; radar at night or restricted visibility in conjunction with an ENC; compass headings and bearings; and GPS plotted position with speed over ground in an ENC. Another comment that was made by a mariner was visual awareness to judge the drift of the boat or how the boat is tracking with the wind; this is of great importance when in close proximity to land and obstructions.

Question 5: What information do you find is vital to safe navigation, maintaining track and position?

Vital elements to safe navigation as stated by the mariners questioned are position; heading; speed over ground; echo sounder; compass headings and bearings; position in relation to any obstructions and hazards; relation to other vessels in our immediate area and navigation aids, for example beacons, lights, hazards, etc.

Question 6: How useful are current systems in relaying the information needed to maintain your position and or track?

Currently most ENC systems like MaxSea and Cmap are very useful and are utilised on a daily basis. Mariners found them to be invaluable in setting waypoints, setting tracks and estimating a time of arrival (ETA). However some of the programs can be clumsy by allowing a deckhand to accidentally knock a button and altering the display. Or they could be highly distracting causing the mariner to focus on the display too much and neglect what is happening to the vessel in the real world.

Question 7: Are there any particular kinds of information you use, that you would like to see displayed better?

Elements that mariners would like to see implemented into a ENC include vessel heading, a three dimensional view around the vessel to give reference to land and other obstacles when manoeuvering at close quarters, tidal information built into the chart plot. There are concerns, however, that adding all this additional information will clutter the screen and overwhelm the mariner.

Question 8: Do you think a 3D model of a nautical chart would convey the chart details better than a 2D model (i.e. and electronic or paper chart)?

When asked if a 3D model of a nautical chart might improve their navigation all replied that it would be, especially in close to land and confined waters. They liked the idea that by having a model they could manipulate the view to suit their needs, for example scanning a work area ahead for possible dangers, obstacles and prominent landmarks, especially if the area is unfamiliar to the mariner.

Question 9: Have you ever been in a situation where navigational information could have been improved, to avoid an incident?

Due to the small amount of people interviewed only a few of the mariners had been in incidents whereby navigation information could have been improved, one mariner had a chart plotter plot a position 150 metres off course. While another commented that Radar is only useful when other vessels are outside of the vessel's sea clutter or inside of the Radar's vertical beam, so it is no good at really close quarters.

Further Comments

Other comments raised by the questionnaire were that integrated displays have the potential to seize and hold the navigator's attention. In some ways it is preferable to have separate instrumentation – thus forcing the navigator to move from one instrument to another, maintaining a visual lookout at the same time. An integrated display is best used by a junior officer, relaying information to the master as required. It can be extremely dangerous for a master to concentrate on a single display.

CHAPTER 3

Mechanisms on Graphical Core Variables in the Design of Cartographic 3D City Presentations

Markus Jobst, Jan Eric Kyprianidis and Jürgen Döllner

Research Group 3D Geoinformation, University of Potsdam - Hasso Plattner Institute, Potsdam, Germany

Abstract

Virtual 3D city models are increasingly used in geospatial planning and discussion processes. This is one reason that the effectiveness and expressiveness of the presentation form has to be investigated from a cartographic point of view. This investigation reaches from recording and modelling procedures for 3D city models to a semiotic model, which has to be adapted for the use of 3D. This contribution focuses on mechanisms on graphical core variables that play an important role for the precise geospatial information transmission with 3D. Methods of non-photorealistic rendering can be combined with cartographic requirements. By this means a new potential for the graphic design of cartographic 3D city presentations can be shown.

Keywords: virtual 3D, non-photorealistic rendering, effectiveness, expressiveness

1 Introduction

Cartographic 3D visualization and virtual 3D worlds are increasingly importance in the communication process of space. By now, tools like WorldWind, Virtual Earth or Google Earth form important geospatial presentation methods in planning and discussion procedures. These tools support naïve geography and thus may enhance the specific transmission of geospatial information (Däßler 2002). In addition communication theories support the assumption that 3D presentation forms establish a naïve information transmission for topographic issues (Egenhofer 1995).

The aspect of successful communication leads to perceptional and graphical design issues. If graphics are used for cartographic depictions, a clear perception and understanding of information content has to be ensured. This demand is also valid for dynamic 3D presentation methods, if these are used for cartographic tasks. Hence an extension of the traditional semiotic model or at least the influence of 3D mechanisms on the semiotic model of 2D cartography (Bertin 1982) has to be considered. Especially cartographic 3D city presentations with their massive occurrence of occlusion call for appropriate dedication of graphical values and design mechanisms instead of photorealistic depictions that often lead to areas of dead information pixels (depending on the resolution of transmitting interface). Whereas photorealistic renderings of virtual 3D city models occupy most of the graphical values, non-photorealistic approaches try to use variables and design mechanisms according to specific visualization tasks or uses.

This contribution focuses on design mechanisms in virtual 3D city presentations, which make use of or influence graphical core variables. In order to enhance specific parts of 3D elements, to support psychological depth cues on Pseudo3D interfaces (computer displays, projections, etc.) or to follow emerging needs to fit in certain statements, these design mechanisms have to be adopted in a controlled way. Some concepts of virtual 3D city presentations will give an idea of computer graphics requirements and possible use cases for the overall composition design. Its possibilities lead to the limitations of photorealistic renderings for cartographic information transmission. In addition the concepts also depend on the design procedures of virtual 3D cities, which result in various detailed reconstructions and therefore offer a usable composition design. For example, building areas that result from the intrusion of streets will neither be applicable for building-wise textures nor any building-wise information attachment. The section of critical amount of graphical variables and design mechanisms uses a general structure of 3D-semiotics and describes the interplay of graphical variables and 3D design mechanisms. For example, lighting influences graphical variables of colour and brightness. Approaches of non-photorealistic rendering (NPR) can then express methods of NPR according to its space modification, give several pictorial examples and describe main advantages. At last we can combine the approaches of non-photorealistic rendering with graphical variables and design mechanisms in 3D in order to show the potential for the design of cartographic 3D city presentations. The obvious surplus value of NPR for virtual 3D city models with cartographic design leads to future work in cartographic 3D city presentations.

2 Concepts of Virtual 3D City Presentations

The visualization concepts of virtual 3D city presentations use photorealistic and non-photorealistic methods. Photorealistic methods try to reconstruct reality in virtual space. Therefore several graphical variables are used for this simulation procedure and little potential is left for additional information coding. Additionally the high resolution and detailed nuances are often not perceptible and have to be aggregated anyhow. In opposite to that, non-photorealistic rendering, like cartoon rendering, offers precise control over colours, sizes, brightness, forms, etc. for appropriate information coding at various levels. The following points show up the main limitations of photorealistic methods for virtual 3D city models.

Photorealistic visualization implies a number of limitations with respect to virtual 3D city models:

1. To produce convincing photorealistic depictions, complete data in high quality have to be processed, e.g., exactly matching façade textures and high-resolution aerial photography. The larger the virtual 3D city model, the higher the costs for data acquisition. In most cases, required data will not be available for a whole 3D city model. As a consequence, the images are faced by a breach of graphics style.
2. To incorporate thematic information (e.g., state of repair, average rental fees) into photorealistic depictions, the information needs to be visually combined with the virtual 3D city model. This turns out to be difficult, because textured façades, roofs, and road systems dominate the image space.
3. To visualize complex information, photorealistic details increasingly interfere with a growing number of information layers.
4. To express objects of city models in different states, e.g., existing, removed, and planned buildings, photorealism does not offer a broad

range of graphics styles to communicate these variations such as by sketchy and outlined drawings.

5. To generate compact depictions for displays with minimal capacities, e.g., on mobile devices, photorealism frequently fails due to the visual complexity inherent in photorealistic images. For example, a scaled-down version of a digital photography has a drastically lower information level compared to a scaled-down version of a hand-drawn sketch of the same scenery.

These drawbacks and limitations are the main arguments for non-photorealistic rendering, which needs specific requirements of the virtual model. On one hand geometry should follow a clear structure/outline for a most expressive cartoon rendering and on the other hand an object-oriented data structure splits up the element in smaller parts that are individually render- and combinable. These main requirements for non-photorealistic rendering lead to the various design procedures for virtual 3D city models.

3 Design Procedures for Virtual 3D Cities

The procedures to create virtual 3D cities vary from precise photogrammetry to block model extrusion and smart city algorithms. Each procedure results in a specific data model that allows for cartographic information preparation. Cartographic information preparation in this context of visualization focuses on the clear visibility on transmitting media, which enables unmistakable perception of the content. It describes the mightiness of data model to be adapted for visualization and thus mutated in generalization processes.

Main production processes for virtual 3D city models span various methods of airborne surveying and photogrammetry, terrestrial surveying with ground plan extrusion or smart city algorithms. Various methods of airborne surveying and photogrammetry make use of high detailed remote sensing, aerial photographs and laser scanning (LIDAR). Detailed *remote sensing*, if the resolution of satellite pictures is high enough, and *aerial photography* produce top view pictures that form a starting material for the analysis of ground plan. In addition stereoscopic methods allow extracting heights and therefore detailed structures of building roofs (Kraus 2007). Because a manual evaluation with its pointwise recording is highly time consuming, semi-automated procedures were developed and can support the modelling of buildings and roofs (Kraus 2007). As a result a detailed and precise structure of buildings including roofs is offered, which is sufficient for flythroughs in very large scales. For its cartographic usage these data mostly have to be simplified, aggregated and their most important characteristics highlighted on most transmitting media. This has to be done because this highly detailed information cannot be successfully transmitted with most scales of use in combination with low media resolution (displays). *Airborne laser scanning* delivers data with similar precision. But the general result, without any postprocessing, is a point cloud of single measurements in form of the regular raster of the laser scanner (Kraus 2007). A first postprocessing leads to a covering hull of the recorded surface that does not include any objectwise information than the height. At least filtering and further postprocessing lead to a covering hull of single buildings that can be enhanced with further (meta)information. Similar to the results of aerial photogrammetry, laser scanning also requires generalization for most of transmitting media because of its detail richness.

Terrestrial surveying with ground plan extrusion uses building plots and the average height of buildings to reconstruct a building block by extrusion of its outline. This procedure cannot reconstruct the roofs. Therefore the virtual 3D city model resulting from this technique is a simple block model, which may be sufficient for many applications and analysis of districts at a medium scale level. Depending on the outline detail of buildings, simplifications have to be done to remove imperceptible details (Kada 2005). The extrusion process is done automatically by delivered data (outline, height). Additionally, for reason of aesthetics synthetic roofs can be added by algorithms. These synthetic roofs will visualize main characteristics but not represent a reconstruction of real world objects.

Smart city algorithms use cell-based clustering to generate generalized 3D city models (Fig 3.1 shows an example of a generalised virtual 3D city). The clustering is based on the infrastructure network and decomposes city areas into clusters. Therefore the infrastructure network uses an implicit hierarchy of network elements that is used to create generalizations at various scales (Glander and Döllner 2007). As input data the infrastructure network is used for the clustering process. The height of clusters is then calculated by the heights of a 3D city model, which may be the result of laser scanning, aerial photography or similar. The infrastructure network uses weighting of its vectors to build up categories and thus follow a desired degree of generalization (DOG). According to a chosen weight and its selected polylines a cell structure can be computed (Glander and Döllner 2007). By cartographic means this resulting cell structure of a virtual city is applicable for small scale applications. On account of the underlying 3D city model, which is used for the heights of clusters, and the hierarchical structure of infrastructure network various levels of generalization can be visualized. This characteristic of the smart city method seems to be most promising for cartographic applications that call for transmitting media adapted scales.

Besides building reconstruction at various scale levels the modelling of terrain is another part in topographic reconstruction that has to be considered. Data structures for terrain models span from regular grids to triangulated irregular networks (TIN) and regular grid with breaklines (Pfeifer 2002). The modelling of terrain is lead by reduction of data size and modelling precision. In terms of visualization it is important that the virtual 3D city model and the virtual terrain model do not perceptibly diverge.

Fig. 3.1. A generalized 3D city model (Glander and Döllner 2007, p 1)

The appropriate adaptation of graphical variables and design mechanisms for a cartographic visualization of virtual 3D city models will require an element-based structure that allows for element-based highlighting and information attachment. Furthermore the element-based structure enables scale-based or accuracy-based variations of the visualization. This demand for an element-based data-structure leads to the question, what kind of graphical variables and design mechanisms can be used in a virtual 3D environment? The following part will describe a general structure of 3D-semiotic and explain the interplay of graphical variables and 3D design mechanisms.

4 The Critical Interplay of Graphical Variables and 3D Design Mechanisms

Since cartography was involved in spatial information transmission usable graphical variables were a key factor for the graphical coding of information. By traditional means eight core variables can be identified. These are colour, size, brightness, form, pattern, orientation and position (x,y) of the element (Bertin 1982). Bertin split the position variable into two parts, x and y. In combination with the graphical elements point, line, area, diagram and font, all imaginable coding can be realized. The extension of this traditional 2D semiotic structure with design mechanisms of 3D does not only lead to an extension, but to an extended semiotic structure for 3D (see Fig. 3.2). An extension of Bertin's list, as developed by MacEachren (1995) to include the variables crispness, resolution and transparency, is not sufficient for the field of 3D cartography. An extension for 3D cartography means that all added design mechanisms of 3D massively influence the coding with graphical variables and elements. Therefore a semiotic structure for 3D includes variables of vision, composition and psychological influences. Composition variables consist of global and element-wise 3D design mechanisms, graphical variables and graphical elements. These components will have a mutual impact on each other.

The original graphical variables of position are not identified as real 3D variables anymore, because elements in the virtual 3D environment are almost fixed to their position. If an element changes its 3D position, then the relation to other elements is heavily influenced. In addition the cognitive impact of the 3D image will change, as perspective impressions and height relations will change with an element's position. For elements that are located in the 2D presentation plane and are not related with perspective consequences, variables of position in this 2D presentation plane are still valid. For example annotations in the 2D presentation plane will have to change their position in x,y picture space in order to stay perceptible.

The semiotic structure for 3D results in a confrontation of graphical variables and design mechanisms of 3D. As the core elements of graphical composition in 2D lead to a visual clear result on the transmitting media and their mutual impact in 2D is well defined (Bertin 1982, Hake and Grünreich 1994, Spiess 1996, Brunner 2001), a dynamic 3D map with its demand of intuitive geospatial information transfer, expands the semiotic model and shows up new graphical conflicts, which are described by Table 3.1 below.

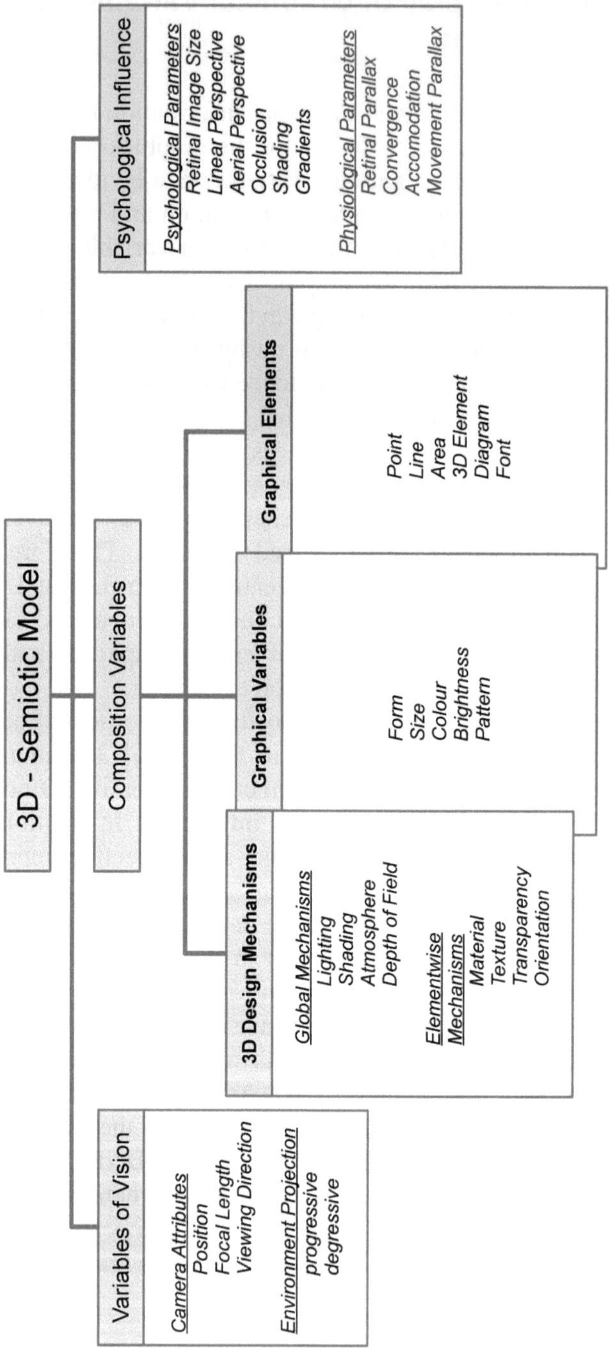

Fig. 3.2. Extended semiotic structure for 3D

The critical interplay of graphical variables and design mechanisms show that composition variables have to be precisely considered for cartographic 3D applications in order to find appropriate graphical expressions. However composition variables in 3D offer additional design mechanisms. The rendering of the virtual 3D environment provides relevant tools to incorporate and manipulate most graphical variables and design mechanisms. These rendering techniques are called non-photorealistic rendering (NPR).

Table 3.1. New graphical conflicts of 3D design mechanisms with graphical variables

	A.Form	B.Size	C.Colour	D.Brightness	E.Pattern
Global Mechanisms:					
1.Lighting			C1	D1	
2.Shading	A2		C2	D2	E2
3.Atmosphere			C3	D3	
4.Depth of Field		B4			
Elementwise Mechanisms:					
5.Material	A5	B5	C5	D5	E5
6.Texture					E6
7.Transparency			C7	D7	E7
8.Orientation	A8				

A2: Shading influences an element's form, if for example hard shading makes it hard or impossible to differentiate between element and shadow. This situation becomes definitely urgent when shadows make use of whole grey values.

A5: Material influences form especially if “bump-mapping”- or “displacement mapping” techniques are used. These techniques apparently deform an element's geometry in order to simulate material surfaces.

A8: Orientation influences form because a perceptible form depends on an element's silhouette. If orientation changes, also the silhouette will change. For specific cases it will be needful to align elements according to their main characteristic. For example the front view of a church.

B4: Depth of field influences perception of size, because growing fuzziness prevents from concrete estimation, although the human cognitive system tries to come to an result by indications (linear perspective, gradients, etc).

B5: Material influences size, if “bump-mapping”- or “displacement mapping” techniques are used. These techniques apparently deform element's geometry in order to simulate material surfaces and therefore may change the size of elements.

C1: Lighting influences colour in context with brightness distribution. The reaction of element's surface with lighting and the combination of lighting colour with element's surface describe further influences. For example the distribution of brightness directly changes the saturation of colour: the brighter an illuminated colour area is, the more unsaturated the colour will appear.

C2: Shading influences colour as saturation and brightness of an element's colour will be changed.

C3: Atmosphere influences colour with its simulation of haze, which changes brightness and saturation of colours.

C5: Material and surface texture influences colour if texture overlays and texture combinations create a new one. For example, a single coloured ball in red that becomes combined with a mottling texture will change saturation and brightness of colour according to the mottling pattern.

C7: Transparency influences colour if it attenuates the opaque characteristic. Overlays with other elements become established and additionally change the allocation of colours.

D1: Lighting directly influences brightness as this is the main characteristic of light.

D2: Shading directly influences brightness as the change of grey values is how shadow works.

D3: Atmosphere influences brightness as the change of brightness is one part of haze simulation.

D5: Material and surface texture influences brightness when combinations of textures result in a new brightness.

D7: Transparency influences brightness as transparency changes reflexivity of the element's surface.

E2: Shading influences pattern if the forming and impact of shadow results in similar grey values as the pattern. Similar grey values then unite and create a new pattern.

E5: Material influences pattern when "bump-mapping" or "displacement mapping" apparently change the element's geometry. Therefore surface shadows are created and thus change original pattern.

E6: Texture influences pattern if texture combinations result in a new pattern.

E7: Transparency influences pattern, because new combinations become enabled with a partial transparent pattern.

5 Approaches of Non-photorealistic Rendering

Non-photorealistic computer graphics denotes the class of depictions that reflect true or imaginary scenes using stylistic elements such as shape, structure, colour, light, shading, and shadowing that are different from those elements found in photographic images or those elements underlying the human perception of visual reality (Figure 3.3 shows an example). Most researchers agree that the term "non-photorealistic" is not satisfying because neither the notion of realism itself nor its complement, the non-photorealism, is clearly defined. Nevertheless, "non-photorealism" (NPR) has established itself as a key category and discipline in computer graphics starting around 1990. An introduction to concepts and algorithms of non-photorealistic computer graphics can be found in Strothotte and Schlechtweg (2002) as well as in Gooch and Gooch (2001).

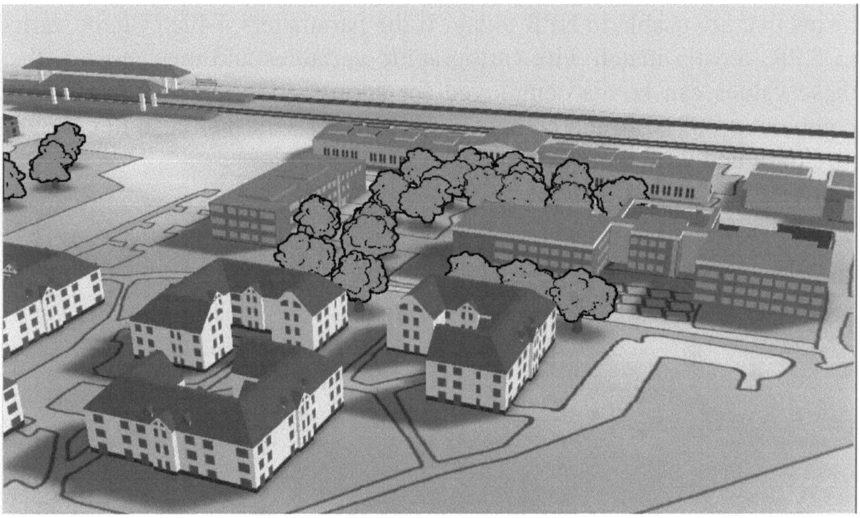

Fig. 3.3. Example for an expressive rendering of a 3D city model

5.1 A Technical Aspect of Non-photorealistic Rendering

Technically, non-photorealistic 3D rendering is implemented by redefining the geometry stage and rasterization stage of the standard 3D rendering pipeline using application-defined procedures, known as shaders, which implement geometric transformations, illumination calculations, and pixel shading. For example, the classical Phong illumination model (Phong 1975) can be substituted by a cartoon-like illumination shader, which reduces the number of shades per colour. Furthermore, non-photorealistic 3D rendering takes advantage of multi-pass rendering, that is, they process a 3D scene description several times, generating and combining intermediate images into a final image. This way, for example, enhanced edges and shadows can be included in the overall rendering process. Furthermore, today's programmable computer graphics hardware supports the implementation of non-photorealistic 3D rendering. Therefore, most techniques can be used in interactive or even real-time applications and systems.

5.2 Combining Graphical Variables, Design Mechanisms and NPR

It seems obvious that non-photorealistic rendering with its cartoon-like visualization directly uses graphical variables and design mechanisms as these are used in cartography. Following descriptions list actual mechanisms that are usable in NPR today. If the parameters, which can be varied in NPR, mostly match with cartographic variables and mechanisms, then these values can be also employed for cartographic aims (Fig. 3.4). The combination of NPR and composition variables will then lead to cartography-designed virtual environments.

Fig. 3.4. Example for usage of NPR and Level of Detail adaptation (Döllner and Walther 2003, p 6)

Generally characteristics of non-photorealistic 3D rendering include the ability to sketch geometric objects and scenes, to reduce visual complexity of images, as well as to imitate and extend classical depiction techniques known from scientific and cartographic illustrations. Fundamental techniques of real-time non-photorealistic 3D rendering address the following characteristics:

1. Lighting and Shadowing. The programmable rendering pipeline allows developers to implement new illumination models such as the model introduced by Gooch et al. (1998). In addition, real-time shadow techniques can be seamlessly integrated and, thereby, provide a valuable depth cue;
2. Colouring and Shading. Non-photorealism allows for vivid and domain-specific colour schemes. Cartoon-like appearance is achieved by specialized colour schemes such as tone shading;
3. Edges and Silhouettes. Edges as visually important elements can be treated as "first-class" objects in depictions, that is, they can be

enhanced and stylized. Isenberg et al. (2003) give an overview of algorithms for outlines and silhouettes; (Fig. 3.5)

4. Texturing. Texturing is used as a fundamental computer graphics operation (Haeberli 1990) to outline the principle curvature of surfaces, to simulate strokes using virtual brushes, to procedurally generate textures, and to implement image-based operations such as image overlays, image blending, convolution filtering, etc.

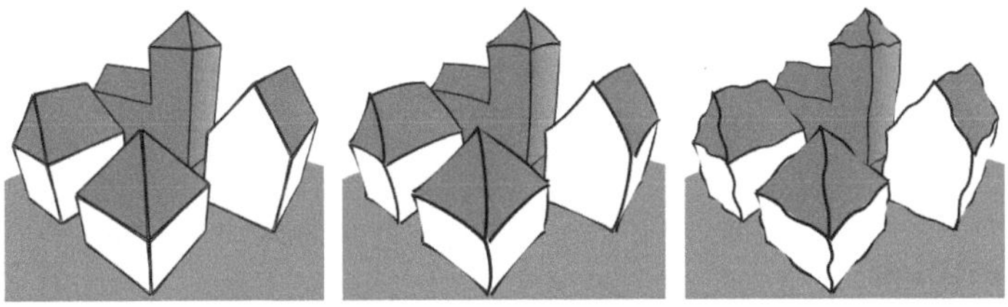

Fig. 3.5. Variations of edge visualization showing the potential of edge rendering in NPR

This section can show that rendering mechanisms in NPR use the same composition variables and designing procedures as in cartography. Therefore these parameters can easily be adapted for cartographic needs. This means that geospatial information description can follow cartographic rules in order to result in an unmistakable information transmission.

6 Conclusions

This contribution focuses on mechanisms on graphical core variables in the design of cartographic 3D city presentations. This means that specific design mechanisms exist for the cartographic visualization of virtual 3D cities. The notion is that NPR describes these methods within the rendering process. Before these visualization methods can be adapted, appropriate virtual 3D city models have to be created in order to lead to a desired result. Unsuitable models may lead to unplanned artefacts in the rendering process that will represent non-important information. On that score recording methods (remote sensing, aerial and terrestrial measurements, etc.) and their postprocessing have to be considered with the aim to use appropriate starting materials for cartographic visualization. Further graphical designing steps for cartographic 3D city presentations lead to the question: can the design mechanisms of 3D can be incorporated in the traditional semiotic model of cartography? Section 4 describes the critical interplay of graphical variables and 3D design mechanisms and results in an extended semiotic structure that is applicable for 3D. Because of the interplay of graphical variables and design mechanisms, conflicts can be identified and are shortly described in that section. The presented methods of non-photorealistic rendering in context with cartographic visualization show their close relation with composition variables. This clarifies the usage of NPR for cartographic visualization of virtual 3D cities.

Future research in this field has to focus on the usability of the extended semiotic model for 3D and the power of variables offered by NPR. In the end usability studies and user tests will show if these techniques can produce expressive and effective results for cartographic 3D city presentations.

References

Bertin J (1982) Graphische Darstellungen – Graphische Verarbeitung von Informationen, übersetzt und bearbeitet von Wolfgang Scharfe. Walter de Gruyter, Berlin

Brunner K (2001) Kartengestaltung für elektronische Bildanzeigen. In: Kartographische Bausteine, Band 19. Technische Universität Dresden

Däßler R (2002) Visuelle Kommunikation mit Karten. In: Engelbert A, Herlt M (eds) Updates – visuelle Medienkompetenz. Königshausen & Neumann, ISBN: 978-3826023699

Döllner J, Buchholz H (2005) Expressive Virtual 3D City Models. Proceedings of the 22nd International Cartographic Conference, A Coruña, Spain

Döllner J, Walther M (2003) Real-Time Expressive Rendering of City Models. Proceedings of the IEEE Information Visualization, pp 245-250

Egenhofer M, Mark D (1995) Naive Geography. COSIT 95, Semmering, Austria. In: Frank A, Kuhn W (eds) Lecture Notes in Computer Science, vol 988. Springer-Verlag, pp 1-15

Glander T, Döllner J (2007) Cell-Based Generalization of 3D Building Groups with Outlier Management. ACMGIS 2007, Seattle, ISBN 978-1-59593-914-2/07/11

Gooch A, Gooch B (2001) Non-Photorealistic Rendering. A K Peters Ltd

Gooch A, Gooch B, Shirley P, Cohen E (1998) A Non-Photorealistic Lighting Model for Automatic Technical Illustration. Computer Graphics, Proceedings of SIGGRAPH '98, pp 447-452

Hake G, Grünreich D (1994) Kartographie, 7. Auflage. Walter de Gruyter Verlag, Berlin/ New York

Haeberli P (1990) Paint by Numbers: Abstract Image Representations. Computer Graphics, Proceedings of SIGGRAPH'90, pp 207-214

Isenberg T, Freudenberg B, Halper N, Schlechtweg S, Strothotte T (2003) A Developer's Guide to Silhouette Algorithms for Polygonal Models. IEEE Computer Graphics and Applications, 23(4):28-37

Kada M (2005) 3D Building Generalisation. Proceedings of the 22nd International Cartographic Conference, A Coruña, Spain

Kraus K (2007) Photogrammetry - Geometry from Images and Laser Scans, 2nd edn. Walter de Gruyter, Berlin, ISBN: 978-3-11-019007-6

MacEachren AM (1995) How Maps Work: Representation, Visualization and Design. Guilford Press, New York

Nienhaus M (2005) Real-Time Non-Photorealistic Rendering Techniques for Illustrating 3D Scenes and their Dynamics. Dissertation, Universität Potsdam, Potsdam

Pfeifer N (2002) 3D Terrain Models on the Basis of a Triangulation. Geowissenschaftliche Mitteilungen, Heft 65, ISBN: 3-9500791-7-3

Phong BT (1975) Illumination for computer generated pictures. Commun. ACM 18:6:311-317

Spiess E (1996) Attraktive Karten – Ein Plädoyer für gute Kartengraphik. In: Schweizerische Gesellschaft für Kartographie SGK (eds) Kartographie im Umbruch – neue Herausforderungen, neue Technologien. Kartographische Publikationsreihe 14:56–69

Strothotte T, Schlechtweg S (2002) Non-Photorealistic Computer Graphics: Modeling, Rendering and Animation. Morgan Kaufman

CHAPTER 4

An Egocentric Urban Viewshed: A Method for Landmark Visibility Mapping for Pedestrian Location Based Services

Phil Bartie[1], **Steven Mills**[1] **and Simon Kingham**[2]

[1]Geospatial Research Centre (NZ), Christchurch, New Zealand

[2]Department of Geography, University of Canterbury, Christchurch, New Zealand

Abstract

A variety of information can be provided to pedestrians using location based services in support of tasks such as wayfinding. Typically current location aware systems use proximity to filter databases for contextual information. We show that a filter based on the visibility of features is a useful additional capability made possible through the use of digital surface models. A number of visibility metrics are suggested for adoption by a location based service, to provide quantitative visibility information so that items of interest may be ranked according to a meaningful priority. Real world experiences validate the usefulness of these metrics, and a number of improvements are suggested.

Keywords: location-based services; digital surface models; viewsheds; visibility metrics; usefulness test

1 Visibility Modelling and Pedestrian Navigation

There is a growing interest in the development of location based services (LBS) in support of pedestrian activities, both rural and urban (Jiang and Yao 2006). The research presented in this paper is in anticipation of devices able to support natural pedestrian wayfinding and navigation in an urban context. It is argued that the requirements for pedestrian navigation are quite different from that of vehicle navigation. While junctions form a key navigation component for motorists, pedestrians more often use landmarks as cues (Millonig and Schechtner 2007). The urban environment is defined by these landmarks, their organisation, and interrelationships (Fisher-Gewirtzman and Wagner 2003) and there is a strong linkage between what a pedestrian can see and how they comprehend a city. We therefore argue that integrating the capability to use landmark visibility information in a navigational device for urban pedestrian use, requires 'egocentric visibility modelling'.

The term 'viewshed' has existed in landscape architecture since the 1960s (Tandy 1967; Lynch 1976), and has been adopted by many disciplines. A viewshed depicts areas which can be seen from a designated observation point, generated by calculating lines-of-sight (LOS) from that point to all other locations within the study area. Most Geographic Information Systems (GIS) offer the functionality to carry out such calculations (De Smith et al. 2007). Visibility is one of the most commonly used GIS analysis tools (Davidson et al. 1993) with an extensive catalogue of research work, including siting radio masts (De Floriani et al. 1994), locating the most scenic or most hidden routes (Stucky 1998), landscape planning (Fisher 1996), as a weapon surrogate in military exercises (Baer et al. 2005), and in examining spatial openness in built environments (Fisher-Gewirtzman and Wagner 2003).

Visibility studies within GIS require access to a digital terrain dataset for the area of interest. These are usually raster grid datasets, considered to be 2.5 dimensional, recording a single elevation value (z) for any location (x,y). In general terms these are known as Digital Elevation Models (DEM), but may be more specifically referred to as Digital Terrain Models (DTM) if they reflect the elevation values of the bare earth, or Digital Surface Models (DSM) if they capture building and vegetation elevations.

There have been numerous previous studies on urban visibility, these have tended to use the 2 dimensional boundary of buildings to calculate isovists (Tandy 1967; Benedikt 1979; Turner et al. 2001). To more closely model urban visibility it is necessary to source a DSM at high resolution, such that building and vegetation profiles are captured accurately. Light

Detection and Ranging (LiDAR) remote data capture techniques have been shown to be suitable for this in urban studies (Palmer and Shan 2002; Rottensteiner and Briese 2002), and are considered superior for LOS calculations than using community contributed 3D models of inconsistent and questionable accuracy, such as with the Google 3D Warehouse.

This paper discusses visibility modelling techniques, and presents a number of additional metrics that will allow future location based services to report landmark information in a more intuitive manner, facilitating the exploration of the city. The paper draws attention to a number of relevant areas of visibility research, including cumulative visibility. It then explains the line-of-sight algorithm used in this research, and a supporting database architecture to model Features of Interest (FOI). Finally the method is demonstrated in a real world situation.

2 Visibility Analysis

If every terrain cell in a line-of-sight path is considered between an observer and target it is referred to as the 'golden case' (Rana and Morley 2002). Although providing the most accurate results from a terrain model this method is computationally expensive, and therefore much of the previous research has focussed on techniques to reduce the number of calculations by considering only visually important cells. Examples of this include using Triangulated Irregular Networks (TINs) (De Floriani and Magillo 1994), or filtering based on topographic features (Rana and Morley 2002). These essentially look to simplify the terrain complexity, or reduce the number of observer-target pairs considered in viewshed generation.

The 'golden case' may be maintained whilst offering rapid retrieval of visibility details by using a Complete Intervisibility Database (CID) (Caldwell et al. 2003), also referred to as a visibility graph (O'Sullivan and Turner 2001), or visibility matrix (Puppo and Marzano 1997). These approaches store the pre-calculated viewshed results from every possible location in a study region, meaning future users only require a simple lookup to return the stored viewshed result. The computational cost of producing a CID is very high, being an $O(n^2)$ calculation, therefore parallel or grid processing techniques are often used (Llobera et al. 2004).

By storing the complete set of viewsheds a number of additional attributes are available for analysis, such as the Cumulative Visibility (Wheatley 1995), or Visual Magnitude (Llobera 2003). These depict the total number of times a cell can be viewed from elsewhere, indicating highly visible regions. Topographically prominent areas such as ridges and peaks often feature highly, however visually prominent landscapes may not necessarily be topographically prominent, such as the high intervisibility which occurs in valleys (Llobera 2003).

2.1 The Urban Cumulative Viewshed

In the context of rural DTMs, peaks and ridgelines form important spaces which act as barriers, and occupiable vantage points. In urban DSMs these ridgelines correlate to building roofs, and are not generally occupiable. This has a number of implications with regard to previous research focussed on reducing observer-target pairs, which consider ridgelines to be significant vantage points (e.g. Rana and Morley 2002). To faithfully replicate the situation in urban space, a cumulative visibility map must be based on a DSM and restrict observer locations to those areas accessible by pedestrians.

Fig. 4.1 shows cumulative visibility for a section of Christchurch, New Zealand. This was produced by calculating viewsheds from 20,000 locations selected at random from all publicly accessible spaces, and summing the results to indicate how many times a cell can be seen. The DSM was created at 1 metre resolution from LiDAR return information. To reduce the impact of edge-effects the analysis was carried out over a larger extent than that shown (Llobera 2003).

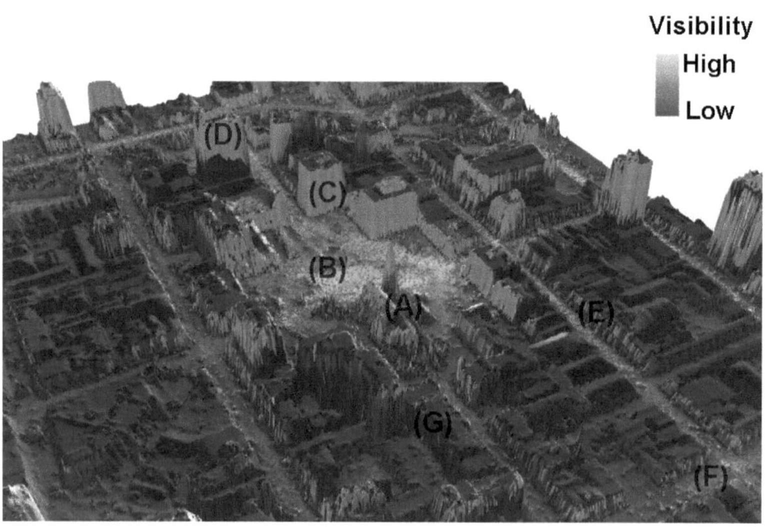

Fig. 4.1. Cumulative Visibility Map for Christchurch, New Zealand

A number of observations can be made:

- Tall buildings have high cumulative visibility values (e.g. Christchurch Cathedral - A)
- Open spaces have high cumulative visibility (e.g. Cathedral Square - B)
- The building frontages adjacent to streets have high visibility values for the entire face (e.g. building C) whilst those faces surrounded by low rise development only receive high scores for the uppermost sections of the face (e.g. building D)
- Low rise buildings only visible from a single street receive mid-range scores (e.g. E)
- Street intersections receive a high score, as they can be viewed from a number of directions (e.g. F)
- The roofs of buildings have low scores as they cannot be viewed from street level (e.g. G).

Depending on the intended purpose, the computational cost of cumulative viewsheds may render them inappropriate. De Floriani (1994) suggested that the cost will not be repaid if there are a minimal number of observation locations. In terms of an LBS where viewshed information is only required for a single user location, it would therefore follow that a cumulative viewshed would not be appropriate. However the cumulative viewshed indicates which areas in an urban scene are important for visibility analysis, and this information could be used in a strategy to reduce the number of target locations for real time visibility analysis conducted on a mobile device.

3 Calculating Egocentric Visibility for LBS

Whilst GISs typically consider the world from above with all areas equally important, an LBS takes an egocentric viewpoint (Meng 2005; Reichenbacher 2005).

A location based service is defined by Jiang and Yao (2006) as an application which is both location-aware, and context-aware, therefore requiring information on the user's position and surroundings. Currently LBSs use proximity as a spatial filter to retrieve relevant contextual information, a notable exception is the Edinburgh Augmented Reality System (EARS) (Bartie and Mackaness 2006) which is able to filter information based on the visibility of FOIs from the user's location.

EARS accesses a database of pre-calculated visibility results for 86 FOIs located around the city of Edinburgh. These results were calculated using ESRI ArcInfo and stored in a relational database management system for rapid sub-second retrieval while on location. As the user explores the city, Global Positioning System (GPS) values are used to locate the user, the application reports what can be seen from the current location. There are a number of drawbacks to this approach including:

- the user's height is fixed to 1.74m for all visibility calculations
- minimal quantitative information is available regarding feature visibility
- the system is unable to accommodate user or community contributed FOIs, as viewshed functionality is not provided on board
- any updates to the DSM require all the visibility calculations to be re-run.

An LBS able to provide the user with information about the visibility of FOIs would be able to guide the user by referring to landmarks (Michon and Denis 2001; Raubal and Winter 2002; Goodman et al. 2004; Ross et al. 2004; May et al. 2005; Millonig and Schechtner 2007), or inform the user about the current surroundings in a natural way. It is therefore necessary to provide an LOS algorithm for use in an LBS which can provide real time quantitative information on the visibility of FOIs.

4 Visibility Implementation for LBS

The visibility algorithms implemented in GISs are often the subject of debate; Fisher (1991) reported that different packages gave significantly different results. Source code for Open Source GIS applications are in the public domain available for scrutiny, whilst the algorithm implementations of commercial software is unknown. A useful survey of the visibility functionality in GIS can be found in a publicly available report to the US Army Line of Sight Technical Working Group (US Army Corps of Engineers 2004).

Riggs and Dean (2007) showed through field trials that predicted viewsheds and surveyed results had lower discrepancies when using higher resolution DSMs. For urban studies using LiDAR datasets it is hoped that discrepancies will be small, although it is acknowledged that a DSM is a 2.5D dataset, and will not report true visibility values under bridges, overpasses, or under vegetation canopy.

Llobera (2003) introduced the concept of 'visual exposure', and suggested that this dynamic aspect of visibility had been overlooked within previous research. Visual exposure focuses on how much of a feature can be viewed from the surrounding space, enabling the creation of surfaces to show in which direction a viewer would need to move to view the target more, or less, clearly. This technique can be used to find visual corridors, or visual ridges, and forms a useful basis for considering LOS in the context of LBS.

For this research a toolkit was written to allow experimentation with the 'golden case' LOS algorithm. The source datasets were a surface model, an observer location, and a database of feature locations. For this study the test area selected was the city of Christchurch, New Zealand. All calculations were carried out using New Zealand Map Grid (NZMG), with the facility for a user height to be specified. LiDAR data was sourced from the Christchurch City Council, and a DSM rendered at 1 metre resolution. For simplicity vegetation was treated as a visual barrier, although the concept of partially obscured views through vegetation has been examined (Llobera 2007).

4.1 The Database Model

A database model was designed such that each FOI entity could be divided into component parts. For example Christchurch Cathedral (Fig. 4.2) could be divided into three parts to represent the spire, the main building, and café annex. Each FOI part could be assigned a number of target locations so the visually important aspects of the structure could be explicitly modelled. It is necessary to place targets around the base of FOIs so that as higher targets are obscured by the building's walls on approach, the LBS does not consider the FOI to be out of view (Fig. 4.3). It therefore follows that a greater proportion of the targets on an FOI should be visible as the observer moves away, although the target will occupy a smaller part of the field of view.

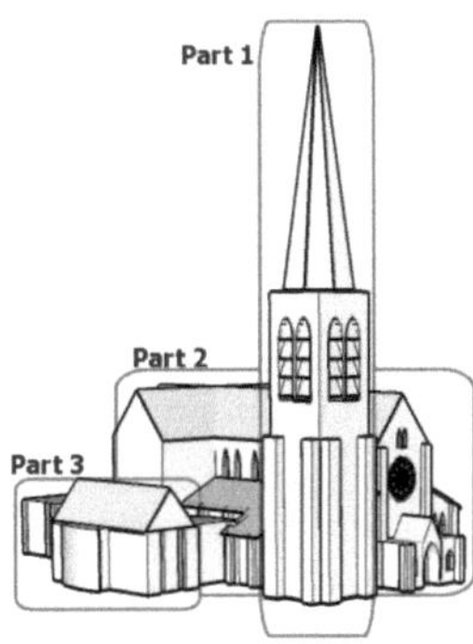

Part 1 – Spire
Part 2 – Main body
Part 3 – Café

Fig. 4.2. Christchurch Cathedral, New Zealand (3D model by ZNO, sourced from Google 3D Warehouse)

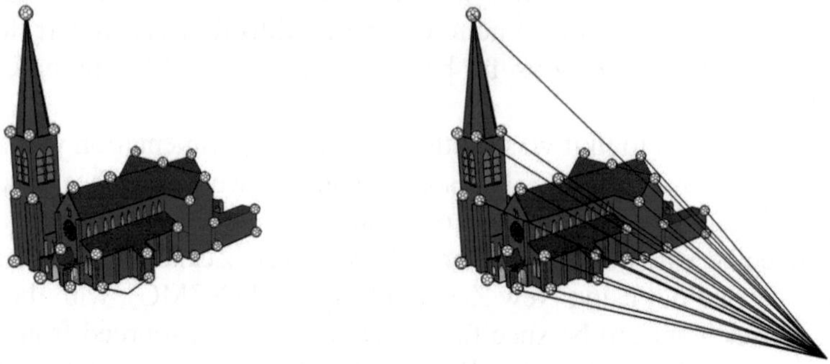

Fig. 4.3. Target Placement (3D model by ZNO, sourced from Google 3D Warehouse)

Provision was made within the database structure (Fig. 4.4) to scan targets according to a visual significance hierarchy. This allowed the LBS to perform a number of scans at varying target densities, firstly to detect the visibility of FOIs, secondly to quantify the visibility information. For this research performance was not a primary consideration, therefore all 2568 cells within the building boundary were used as target locations, at a resolution of 1 target per square metre.

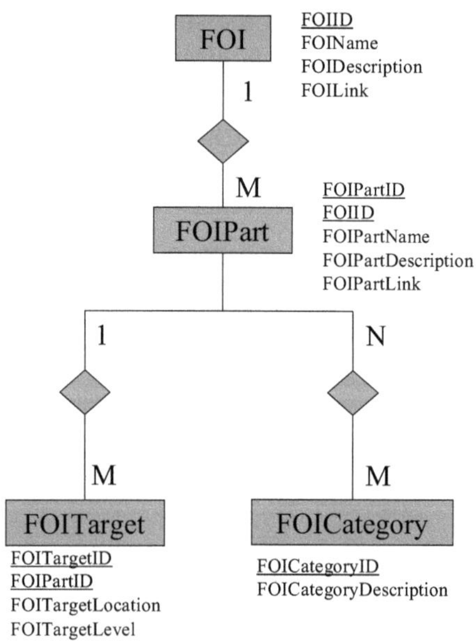

Fig. 4.4. Entity Relationship Diagram for FOI Database

The allocation of category classes at FOI Part level allows for aspects of an FOI to be excluded or included in the results, depending on the user's preferences. Each FOI Part could belong to more than one category class.

4.2 Line of Sight Metrics

When exploring the urban environment a pedestrian's view is filled with features competing for attention. Some distant FOIs may be clearly visible, yet close items may be partially obscured. If these qualities are to be modelled then a set of corresponding metrics are required. These include a metric for proximity to the object, a metric for the amount of a feature that is visible, the field of view occupied, and to indicate if the FOI is on the skyline.

The approach used here was to consider an LOS from the observer to each of the FOI targets in turn, recording the vertical visible extent, and the location of close and distant horizons. These results are then combined to form a number of metrics for each FOI part. The intention is that future LBS devices could use these values to deliver information to a user in a meaningful order, filtering out details not relevant to the current context.

4.2.1 Close Horizon

The 'Close Horizon' is calculated by locating the feature which creates the steepest viewing angle between observer and target, not including the FOI itself. The elevation of this object, known from the DSM, is used to calculate the intercept of a line of sight with the target, and deduce how much of the target is visible (Fig. 4.5 - *TH1)*. The obscured area is also recorded (Fig. 4.5 - *TH2)*.

4.2.2 Distant Horizon

By extending the LOS ray beyond the target it is possible to discover if the target makes the skyline, or is overshadowed by a more distant object. The search continues until either it intercepts the DSM, or reaches the same elevation as the maximum elevation in the DSM dataset. If the ray intercepts the DSM then the 'Distant Horizon' location is recorded along with the elevation value at that point, and both distance behind target and extent of the object showing *(*Fig. 4.5 - *HD1)* are calculated. If no feature is found the target is designated as being on the skyline.

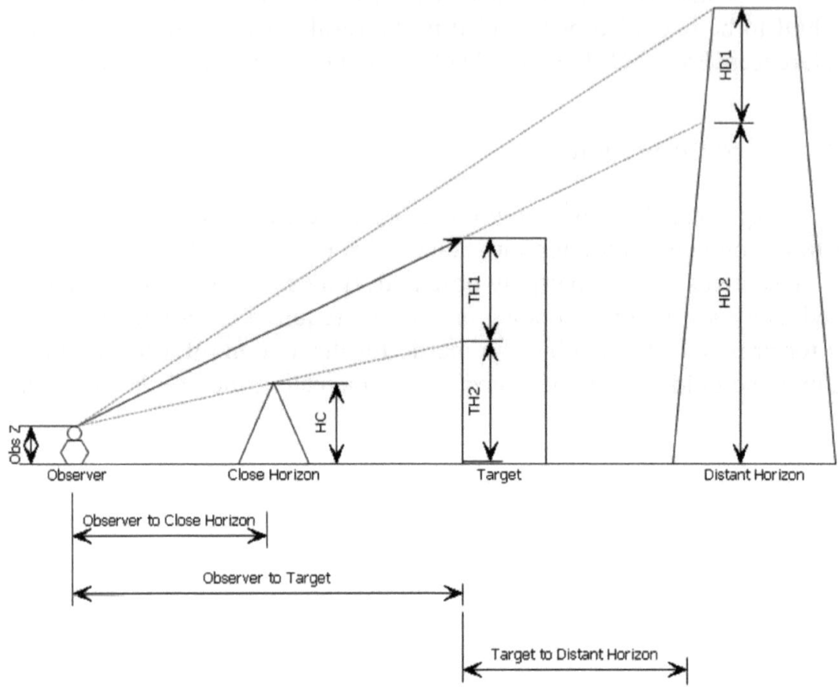

Fig. 4.5. Line of Sight Details; the Close Horizon is used to calculate how much of the target is visible; the Distant Horizon information indicates whether the target is on the skyline or overshadowed by a taller object

The values stored from each of the LOS calculations between the observer and each target point are summarised in Table 4.1.

4.3 Visibility at Feature Level

For an LBS to make use of the target visibility attributes a number of summaries at the FOI Part level are required. These summaries (Table 4.2) are intended to provide an LBS with the facility to filter spatial databases, and to sort results according to various quantitative measures. As the LBS is able to determine automatically which aspects of a feature are visible, it can customize the information delivered. For example it is able to report details of the highly visible spire, but not mention the currently obscured entrance lobby.

Table 4.1. Line of sight target return values

Criteria	Data type	Details
Visible	Boolean (True or False)	Whether a target point can be seen or not from current location
Target Location	Point (x,y)	The location of the target point
Target Elevation *(TH1+TH2)*	Metres (vertically)	The height of the target point
Distance from Observer *(Distance O-T)*	Metres (horizontally)	The distance from observer to target point
Close Horizon Location	Point (x,y)	The location of the tallest object in the line of sight from the observer to the target
Close Horizon Height *(HC)*	Metres (vertically)	The height of the tallest object between observer and target
Close Horizon Intercept With Target *(TH1)*	Metres (vertically)	The amount of target which shows above the tallest near object
On Skyline	Boolean (True or False)	Whether the target has a taller visible object behind it, or sky
Distant Horizon Location	Point (x,y)	The location of the intercept with a taller object behind the target (if any)
Distant Horizon Distance Behind Target *(Distance T-DH)*	Metres (horizontally)	The distance from the target to the horizon
Distant Horizon Elevation *(HD1+HD2)*	Metres (vertically)	The elevation of the item on the horizon visible behind the target
Distant Horizon Intercept *(HD1)*	Metres (vertically)	The amount of the horizon that is visible above the target
Elevation Showing as Ratio Of Distance to Observer Ratio *(TH1 /Distance O-T)*	Ratio	Considering any near horizons, calculate the ratio of visible vertical extent divided by distance to observer

Table 4.2. Visibility metrics at FOI part level

Criteria	Data type	Description
Visible	Boolean (True or False)	Shows if part of a feature can be seen
Average Distance of Visible Targets	Metres (horizontally)	The average distance to only those targets visible to the observer
Maximum Horizontal Field of View	Degrees (horizontally)	The field of view between the widest visible targets on an FOI
Number of Visible Targets	Integer	A count of the number of visible targets
Percentage of Targets Visible	Decimal	Count of targets visible divided by all targets on FOI Part
Total Face Area Visible	Square Metres	The combined total area of feature frontage visible, when considering close horizon
Total Face Area Blocked From View	Square Metres	The area on the frontage which is in the shadow of near blocking objects
Percentage of Targets On Skyline	Ratio	Count of visible targets on skyline, divided by all visible targets
Average Visible Target Height	Metres (vertically)	Average elevation calculated from all visible targets
Minimum Visible Target Height	Metres (vertically)	Lowest visible target elevation
Maximum Visible Target Height	Metres (vertically)	Highest visible target elevation

4.3.1 Field of View

Most of these metrics are self-explanatory, however 'Field of View' and 'Total Face Area' require clarification. The 'Field of View' is calculated between the most extreme targets visible, which make the widest angle. If an object obscures a side of the FOI frontage then the FOV angle will decrease. However if an object blocks a portion in the middle of the FOI with the outside targets still visible, the FOV will return the widest viewing angle calculated from the outside points, ignoring the obscured portion of the FOI frontage.

4.3.2 Total Face Area – Visible / Blocked

The 'Total Face Area' visible is calculated by summing the area visible under each target after considering the area obscured by obstacles between the observer and target, as illustrated in Fig. 4.6. The difference between DTM and DSM establishes FOI height, removing topography from the resulting area. The Total Face Area Blocked reports the area which is shadowed by near objects.

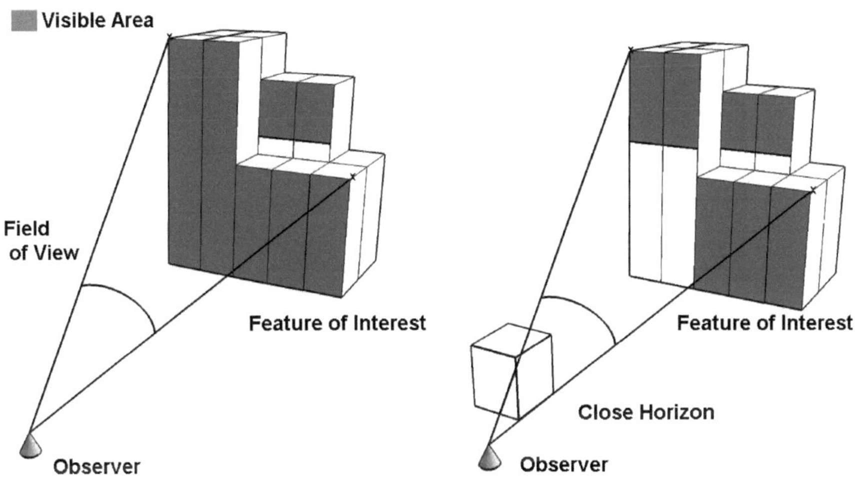

Fig. 4.6. Total Face Area Visible for a Feature of Interest

A number of derived metrics may also be useful such as the ratio of distant horizon height showing (HD1) over distance behind target, to give an indication of dominance of any distant features. Also the total visible face area divided by the average distance to the targets would give an indication of the presence of an FOI from the user's viewpoint.

5 Implementation and Evaluation

An initial demonstration of the method was carried out in the city of Christchurch, New Zealand. Christchurch Cathedral was selected as an FOI, and divided into 3 component parts as outlined in Fig. 4.2. A number of observation points were selected to give different views of the Cathedral, photographs were taken at each site, and the GPS locations passed to the algorithm to return the visibility metrics. Fig. 4.7 shows the location of the test sites, along with corresponding photographs.

Fig. 4.7. Map of Test Sites in Christchurch, New Zealand (LiDAR data sourced from Christchurch City Council)

Table 4.3 shows the results from running the visibility calculations. The third part of the FOI was not visible from either location A or B, and has been removed from the table.

Table 4.3. Results from sites A and B

Location		A		B	
		Part 1	Part 2	Part 1	Part 2
Visible Targets (%)		18.2	0.8	17.7	0.6
Percentage Elevation Showing of the Visible Targets (%)		13.4	8.2	12.9	7.9
Total Face Area Visible (sq m)		202.5	33.4	191.5	28.0
Total Face Area Blocked From View by Near Horizon (sq m)		1325.9	376.1	1316.5	318.7
Average Distance of All Visible Targets (m)		1636.5	1620.5	721.6	706.4
Percentage of Targets on Skyline (%)		3.0	0	2.5	0
Visible Target Height (m)	Average:	41.3	25.6	41.9	26.6
	Maximum:	59.1	29.0	59.1	29.0
	Minimum:	20.1	22.0	26.1	22.0
Maximum Horizontal Field of View for Entire FOI (degrees)		0.25	0.11	0.56	0.24

The majority of the results are as expected with the more distant site showing the FOI to occupy a narrower field of view, that the spire (FOI Part 1) is visible, and none of main building (FOI Part 2) makes the skyline due to the tall surrounding buildings.

However the results show that the percentage elevation of visible targets (13.4% at A, 12.9% at B), percentage of targets on the skyline (3% at A, 2.5% at B), and total face area visible (202m^2 at A, 191m^2 at B) for the spire (FOI Part 1) go against the expected trend and are slightly greater at Location A than B. It is also noticeable that a greater extent of the spire is visible at location A (20.1m to 59.1m at A, and 26.1m to 59.1m at B).

In fact, although counter-intuitive, these values match the real world experience as seen in the photographs taken from these sites. At Location A the vertical extent of the spire visible is greater with the majority of the left side making the skyline, as annotated in Fig. 4.8. At Location B the distant skyscraper blocks the sky behind the spire, and trees in the foreground obscure the lower aspects of the spire. This is in agreement with the output values from the algorithm.

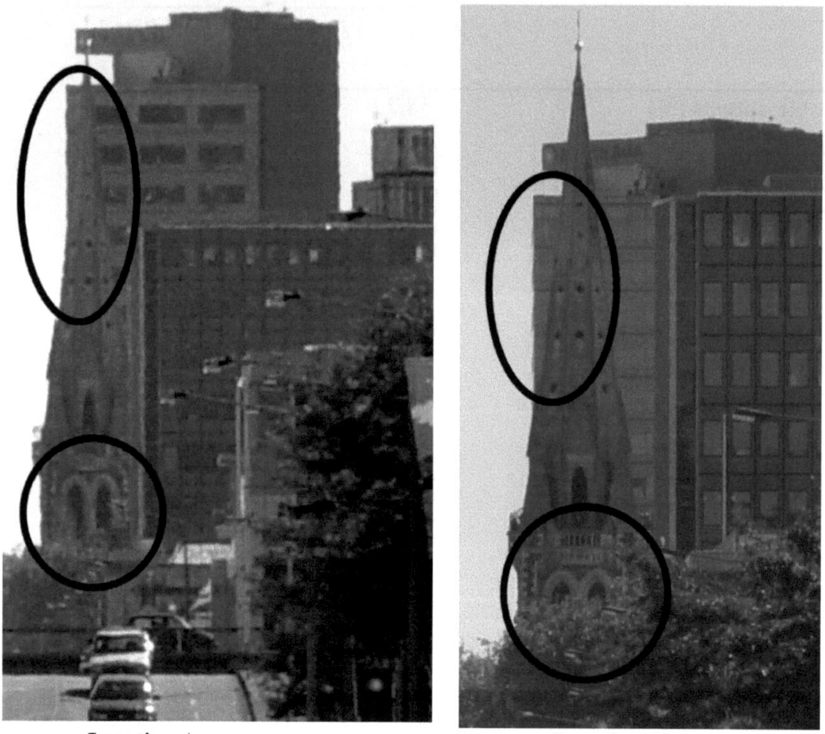

Location A **Location B**

Fig. 4.8. Comparison of Cathedral Spire from Locations A and B

5.1 Additional Analysis

Five further test locations (Fig. 4.9) were selected and the metric results calculated. Values for total visible area frontage along a transect line passing from Location C, through D, and E are shown in Fig. 4.10.

A statue and several trees block the view of the Cathedral along this approach, as reflected by the metrics.

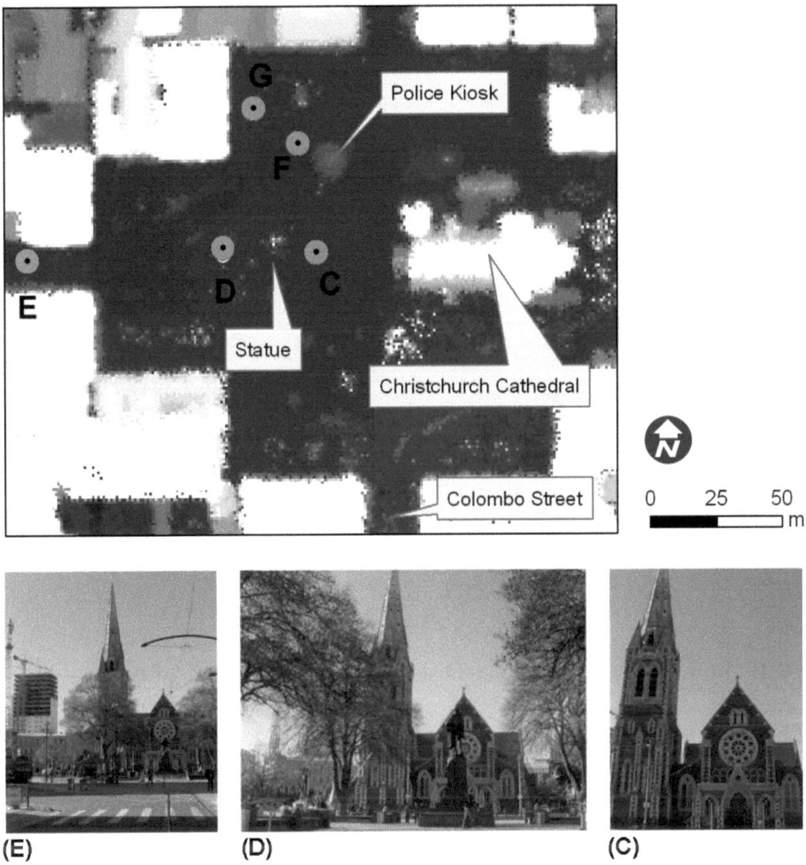

Fig. 4.9. Cathedral Square Test Sites (LiDAR data sourced from Christchurch City Council)

Considering the percentage elevation of targets visible, at Location C 39.9% of the spire is visible, at Location D the main body of the Cathedral receives a score of 17.2%, whilst the spire receives a value of 11.8%, indicating the spire is obscured more than the main body. At Location E the main body received a score of 20.4%, while the spire receives 34.8% indicating the prominence of the spire once more.

The main difference in the view between locations F and G is the proximity of a Police hut (Fig. 4.11). At Location F, the Café (FOI Part 3) is not visible, whilst at Location G it scores 6.5% (percentage elevation of visible targets).

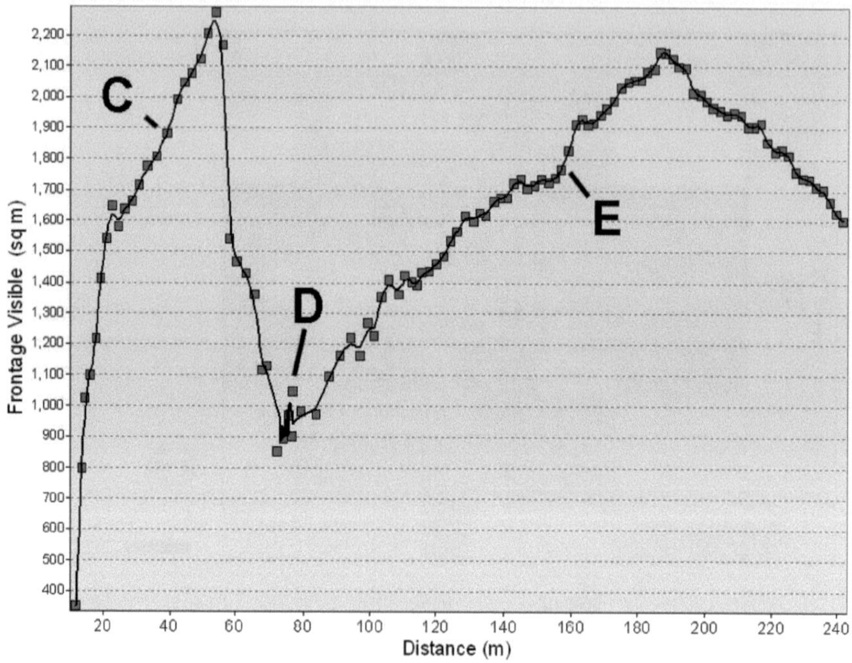

Fig. 4.10. Frontage Area Visible against Distance from Feature of Interest

The total face area of the spire at Location F is 615.1m^2, whilst at Location G it receives a value of 818.3m^2. These are in agreement with the photographic evidence.

Location F

Location G

Fig. 4.11. Site F and G photographs

There is an interesting relationship between the field of view, and total area visible. Fig. 4.12 illustrates this by considering 26,753 locations around Cathedral Square spaced 1 metre apart.

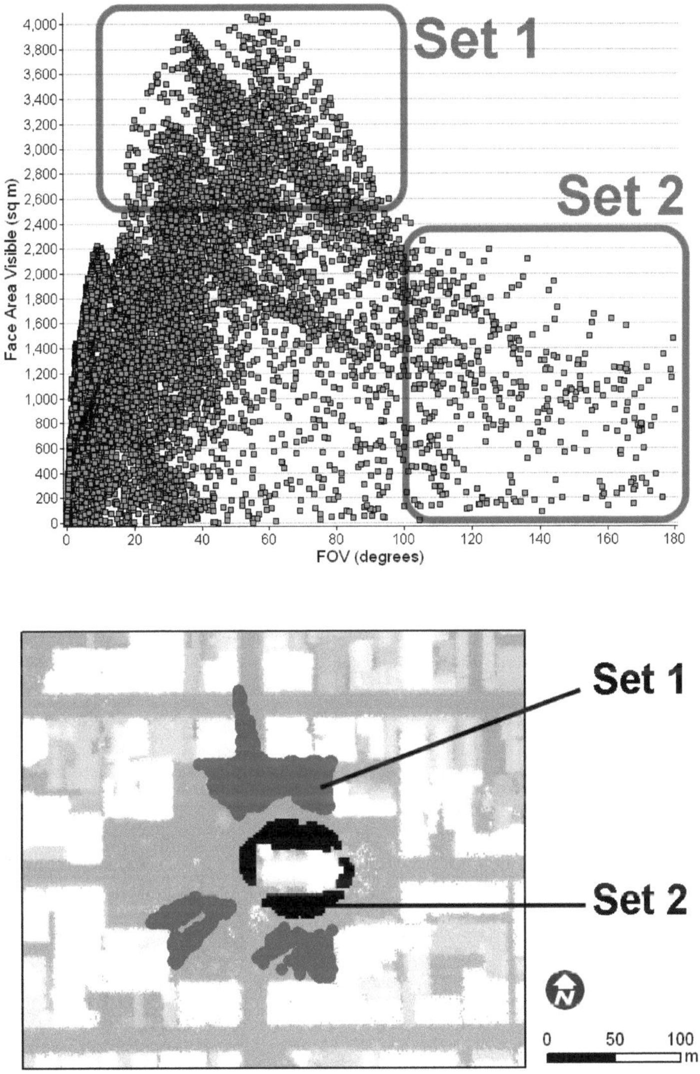

Fig. 4.12. Field of View and Face Area Visible (LiDAR data sourced from Christchurch City Council)

Close to a feature the field of view metric scores high, and the total frontage face area visible is low. As can be seen from the graph the points which contribute to the peak of face area value points (Set 1) are located away from the Cathedral on the edge of the square, whilst the highest FOV values (Set 2) are near the Cathedral. From a viewing experience the face area values may be considered the most appropriate metric to reflect 'how much' of an FOI can be seen, and should be considered with distance to quantify the presence of an FOI.

5.2 Mapping the Visibility Metrics

The values from the LOS implementation may be mapped to indicate how a user's experience of an FOI would vary across space (Fig. 4.13). In this example the area in front of the Cathedral between test locations C and D (Fig. 4.9) enjoys the highest visible percentage of targets, essentially the clearest view.

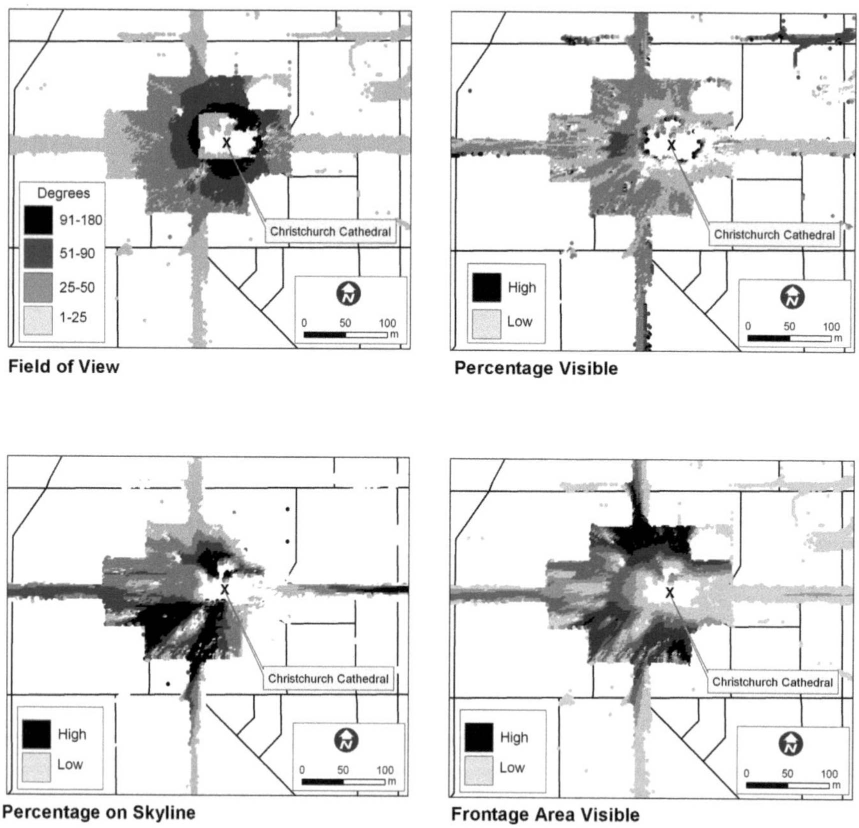

Fig. 4.13. Maps of Key Metrics

Calculating the gradient from any of these datasets will show the magnitude and direction of change of the metric, and may be used to indicate in which direction a user should move to see more, or less, of the FOI.

5.3 Using the Visibility Metrics

These metrics combined with other GIS datasets can provide an LBS with the ability to be context-aware when delivering information to a user. When searching spatial databases the visibility metrics may be used to rank the results showing the most visible items first. In wayfinding, the visible area values may be used to lead the user towards good viewpoints for nominated landmarks. Any information delivered to the user would need to be supported by additional datasets, such that attributes including the FOI's name, address, usage, and history could be conveyed to the pedestrian.

Although not implemented at this stage, a fuzzy logic layer will be introduced in the next phase of the research so that a natural language engine may select the most appropriate English terminology to describe the scene. This will allow the LBS to take on the role of a virtual city guide. The class limits will be set after conducting user trials to evaluate perceived object visibility against the metric results.

To ensure the LBS remains responsive to the user's movements, the metrics must be available in real time. This can be achieved in a number of ways. One method is to pre-cache the results for an area of interest, such that the mobile client requires only a simple lookup of the corresponding FOI visibility summaries.

An alternative approach, which will be used for the next phase of this research, is to implement a client-server architecture whereby a mobile client sends the user's location information across a network to the server which returns the visibility results for the surrounding region. Pre-caching on the server side is also possible such that the most commonly visited areas can be held in memory for improved response times.

6 Conclusions and Future Work

This research has shown that a line of sight algorithm may be used to supply a number of useful contextual visibility metrics from a LiDAR dataset in an urban environment. These results form part of a function of visibility which can be used to prioritise information on features of interest from the current observation location.

We have shown that it is possible to incorporate a wide range of visual statistics into reports of object visibility, with consideration of the surrounding cityscape, and the importance of close and distant horizons.

There are a number of areas for further research in this field. The algorithm considers only the physical aspects of visibility which can be calculated from a DSM. It currently neglects to consider the time of day, or weather in the visibility calculations.

It would also be beneficial to measure the texture, material, colour and contrast differences between the target and any distant horizon objects, such that a metric may be established to indicate how easily an FOI may be resolved from its background. Aerial imagery might offer a partial solution suitable for identifying vegetation backgrounds, but roof top colours will not assist in contrast and colour information from the user's viewpoint, so georeferenced street level photography would be preferable.

Partial visibility through vegetation (Llobera 2007) could be explored such that lines of sight are able to pass through, and under, canopy layers. The algorithm should be extended to accommodate seasonal tree canopy and vegetation density variation. It would also be worthwhile to examine raw LiDAR returns to produce detailed surfaces for the canopy layer.

Currently a scan in the vertical axis over the FOI returns information on the distant horizon. This could also be applied across the horizontal axis so that the surroundings may be used as context to determine if an FOI profile is significantly different from its neighbours (e.g. skyscraper amongst low rise buildings).

Referencing a number of FOIs in the metrics would allow an LBS to use relative descriptions of space, such as " the Chalice monument is visible to the right of the Cathedral". Accessing models from the Google 3D Warehouse may also be useful such that pictorially the shape and texture of FOIs could be displayed, allowing the user to more easily identify the target FOI from surrounding buildings.

The ultimate goal of this research would be a single visibility function which considers weighted metrics based on visibility, architectural interest, building form, and social factors such as building use or related interests to the user, that would allow an LBS to rank in order from a scene those items which a viewer would consider most interesting. The values would then be passed through a fuzzy logic layer and a natural language engine to generate appropriate sentences to convey to the user descriptions of their surroundings, such that the LBS acts as a virtual city guide.

Acknowledgements

The authors would like to thank the Christchurch City Council for use of their LiDAR dataset, and ZNO for the 3D model of Christchurch Cathedral. The research would not have been possible without funding from the University of Canterbury, New Zealand. We would also like to thank Dr William Mackaness for comments on an early draft of the paper.

References

Baer W, Baer N, Powell W, Zografos J (2005) Advances in Terrain Augmented Geometric Pairing Algorithms for Operational Test. Modelling and Simulation Workshop, Las Cruces, NM

Bartie PJ, Mackaness WA (2006) Development of a speech-based augmented reality system to support exploration of cityscape. Transactions in GIS 10:63-86

Benedikt ML (1979) To take hold of space: isovists and isovist fields. Environment and Planning B 6:47-65

Caldwell D, Mineter M, Dowers S (2003) Analysis and Visualisation of Visibility Surfaces. http://www.geocomputation.org/2003/Papers/Caldwell_Paper.pdf (accessed: 2nd November 2007)

Davidson DA, Watson AI, Selman PH (1993) An evaluation of GIS as an aid to the planning of proposed developments in rural areas. Geographical Information Handling: Research and Applications, pp 251–259

De Floriani L, Magillo P (1994) Visibility algorithms on triangulated digital terrain models. International Journal of Geographical Information Systems 8:13-41

De Floriani L, Marzano P, Puppo E (1994) Line-of-sight communication on terrain models. International Journal of Geographical Information Systems 8:329-342

De Smith MJ, Goodchild MF, Longley P (2007) Geospatial Analysis: A Comprehensive Guide to Principles, Techniques and Software Tools. Troubador Publishing

Fisher-Gewirtzman D, Wagner IA (2003) Spatial openness as a practical metric for evaluating built-up environments. Environment and Planning B: Planning and Design 30:37-49

Fisher PF (1991) First experiments in viewshed uncertainty: the accuracy of the viewshed area. Photogrammetric Engineering & Remote Sensing 57:1321-1327

Fisher PF (1996) Extending the applicability of viewsheds in landscape planning. Photogrammetric Engineering and Remote Sensing 62:1297-1302

Goodman J, Gray P, Khammampad K, Brewster S (2004) Using Landmarks to Support Older People in Navigation. 6th International Symposium on Mobile Human-Computer Interaction: MobileHCI 2004, Springer, Glasgow, UK

Jiang B, Yao X (2006) Location-based services and GIS in perspective. Computers, Environment and Urban Systems 30:712-725

Llobera M (2003) Extending GIS-based visual analysis: the concept of visualscapes. International Journal of Geographical Information Science 17:25-48

Llobera M (2007) Modelling visibility through vegetation. International Journal of Geographical Information Science 21:799-810

Llobera M, Wheatley DW, Steele TJM, Cox S, Parchment O (2004) Calculating the inherent visual structure of a landscape ('total viewshed') using high-throughput computing. XXXII International Conference - Computer Applications in Archaeology 2004 - Computer Applications and Quantitative Methods in Archaeology, Beyond the Artifact: Digital Interpretation of the Past, Prato, Italy

Lynch K (1976) Managing the sense of a region. MIT Press Cambridge

May AJ, Ross T, Bayer SH (2005) Incorporating landmarks in driver navigation system design: An overview of results from the REGIONAL project. Journal of Navigation 58:47-65

Meng L (2005) Ego centres of mobile users and egocentric map design. In: Meng L, Zipf A, Reichenbacher T (eds) Map-based Mobile Services. Springer, Berlin

Michon PE, Denis M (2001) When and why are visual landmarks used in giving directions? Spatial Information Theory, Foundations of Geographic Information Science, 292-305

Millonig A, Schechtner K (2007) Developing landmark-based pedestrian-navigation systems. IEEE Transactions on Intelligent Transportation Systems 8:43-49

O'Sullivan D, Turner A (2001) Visibility graphs and landscape visibility analysis. International Journal of Geographical Information Science 15:221-237

Palmer TC, Shan JA (2002) Comparative study on urban visualization using LIDAR data in GIS. URISA Journal 14:19-25

Puppo E, Marzano P (1997) Discrete visibility problems and graph algorithms. International Journal of Geographical Information Science 11:139-161

Rana S, Morley J (2002) Optimising Visibility Analyses Using Topographic Features on the Terrain. Centre for Advanced Spatial Analysis, University College London, London, UK

Raubal M, Winter S (2002) Enriching Wayfinding Instructions with Local Landmarks. In: Egenhofer MJ, Mark DM (eds) Geographic Information Science. Proceedings of the 2nd International Conference, GIScience 2002, Boulder, CO, USA. Springer, Berlin Heidelberg

Reichenbacher T (2005) Adaptive egocentric maps for mobile users. In: Meng L, Zipf A, Reichenbacher T (eds) Map-based Mobile Services. Springer, Berlin

Riggs PD, Dean DJ (2007) An investigation into the causes of errors and inconsistencies in predicted viewsheds. Transactions in GIS 11:175-196

Ross T, May A, Thompson S (2004) The Use of Landmarks in Pedestrian Navigation Instructions and the Effects of Context. 6th International Symposium on Mobile Human-Computer Interaction: MobileHCI 2004. Springer, Glasgow, UK

Rottensteiner F, Briese C (2002) A New Method for Building Extraction in Urban Areas from High-Resolution Lidar Data. International Archives of Photogrammetry and Remote Sensing 34:295-301

Stucky JLD (1998) On applying viewshed analysis for determining least-cost paths on Digital Elevation Models. International Journal of Geographical Information Science 12:891-905

Tandy CRV (1967) The isovist method of landscape survey. Symposium: Methods of Landscape Analysis, pp 9–10

Turner A, Doxa M, O'Sullivan D, Penn A (2001) From isovists to visibility graphs: A methodology for the analysis of architectural space. Environment and Planning B: Planning and Design 28:103-121

US Army Corps of Engineers (2004) Line-of-Sight Compendium. http://www.tec.army.mil/operations/programs/LOS/LOS Compendium.doc

Wheatley D (1995) Cumulative viewshed analysis: a GIS-based method for investigating intervisibility, and its archaeological application. Archaeology and Geographical Information Systems: A European Perspective

CHAPTER 5

A Multi-scale Dynamic Map Using Cartograms to Reflect User Focus

Grant Carroll[1] **and Antoni Moore**[2]

[1]Marlborough District Council, Blenheim, New Zealand

[2]School of Surveying / Department of Information Science, University of Otago, Dunedin, New Zealand

Abstract

A dynamic map that recursively uses a cartogram algorithm in a nested structure to distort shape, enlarging any areas of user interest or focus, has been implemented and tested. The map organises space into a hierarchy of commonly-perceived geographical areas, which is the structure upon which distortion takes place (island at the country level, provinces at the regional level, districts, cities and suburbs). This approach is proposed to be more usable than traditional methods of multi-scale representation or moving between scales. The advantage of using this system is that the local view is expanded, whilst still allowing the global view to be maintained. Although akin to the fisheye display, the novel use of an automated cartogram (with area effectively being proportional to user interest) means that distortion calculation occurs to a great extent only where it is needed, to preserve the local egocentric view, whilst harnessing the cartogram's ability to maintain approximate geographical shape.

This new method was tested to see if it was more usable than other traditional methods of statically or dynamically handling multiple scales (zoom in/out or the use of an inset map) using time as a measure. The cartogram map was found to elicit faster times to navigate to specific locations than the other methods, suggesting that it is more efficient. General satisfaction with the map was expressed in a survey, indicating overall a useful and usable way to navigate through space. This is despite radical distortions of shape, tested in an associated country / continent shape recognition test (it was verified elsewhere that the participants had total knowledge of the shapes of landmasses tested) to reveal that shape is the most important variable. Finally, relative orientation was tested with the cartogram map but no significant results were yielded.

Keywords: egocentric view, density-equalising cartogram, hierarchy, shape distortion

1 Introduction

1.1 Research Problem

Digital maps conventionally use a zoom function to allow users to view areas of interest in greater detail. The disadvantage of using such a function is that it removes the user from the full extent and therefore overall context of the map (recognisable features – cognitive cues - are more likely to be seen at smaller scales). As their view of a map becomes more and more localised, they can become increasingly disconnected from the global view. There are several ways of overcoming this; two major ones will be highlighted here as precursors to the technique presented in this paper:

- The use of two displays either side by side or with one as an inset map within the same display (in both, one map shows the local view while the other shows the global view highlighting the zoomed area). The problem with this is that there is no sense as to what is in the neighbouring areas. Also, in the case of the inset map, valuable room in the display is occupied.
- Using simple locally-emphasising projective or fisheye distortion to allocate more space on the map to the area of interest, compressing the space in which surrounding features are displayed. However, these are globally applied and transform space without regard for the features that occupy it (however, a dual locally and globally projected map has been

presented, actually as a solution to the shifting of reader attention associated with the inset map technique - Snyder 1987).

Cartograms (see Figure 5.2 for an example), also known as a value-by-area map (Dent 1985), are central to an alternative method for maintaining a simultaneous local and global map (hereafter known as the cartogram map) which is presented in this paper. They are a useful way of presenting spatially distributed statistical information so that it is possible to distort the size of an areal unit to show the magnitude of the information which is something other than area. Since their first appearance in modern form in the mid 19th century (Tobler 2004) they have been popularly used to depict socio-economic variables such as population, electoral turnout and so on.

1.2 The Proposed Solution

The purpose of this research is to extend the use of the cartogram to retain a simultaneous global and local view on the same map (an attribute shared with the fisheye method). This is essentially an egocentric map display, which is a cognitively attuned display (Olson 1984; Gallistel 1990), increasingly coming to the forefront in the context of personalised geo-services and increasing use of mobile devices and their limited displays (Harrie et al. 2002; Gutwin and Fedak 2004; Meng 2005). Enlarging the amount of local space available at the expense of peripheral space would make the most of the small display area available.

Whilst there has been some use of different map projections to distort the way a map is presented to suit the egocentric view (Hägerstrand 1957; Kadman and Shlomi 1978; Sarkar and Brown 1994; and Yang et al. 2000), there has been no research into taking the principles of cartograms and applying them to a local area to expand it, specifically based on the map user's interest (though some have developed related solutions – Cuff et al. 1984; Snyder 1987; Leung 1989; Harrie et al. 2002). In effect, this proposal would actually combine the two aims of variable-scale maps, as stated by Harrie et al. (2002). These aims are the conventional use of cartograms to make the density of a variable uniform and; adjusting a constant-scale map to create more space for local display.

Hägerstrand's azimuthal projection in particular emphasises the perspective of the individual, spatially biased towards their area of domicile, where they have a greater amount of spatial knowledge. Consequently areas central to the map are exaggerated with this projection, while peripheral areas are minimised in area – the use of the cartogram in this study attempts to achieve the same cognitively-plausible principle, but in a less-structured, though more flexible, geographically-informed and dynamic way. In effect, by being offered as an alternative to the zoom in operation in managing multiple scales, the function of the cartogram has shifted from static display to dynamic display in an interactive context.

Importantly, cartograms focus on distorting the actual spatial features themselves (normally polygons) rather than trying to distort the surface on which the data is presented, as with locally-focused map projections. In other words, the geographical intelligence implicit in the cartogram relating to approximate maintenance of shape indicates a more refined method.

In creating the cartogram map a solution to the problem of loss of context seen with zoom in operations is presented. By preserving all the relevant adjacent data (albeit compressed) and just enlarging those areas of interest (i.e. making distortion only a local operation as opposed to the global transformation of space with the locally-focused map projections), the user can still maintain a reference to the global view of the map.

The concept of the cartogram map is enabled by its digital context. Likewise, the necessary use of automated cartograms (cartogram algorithms) is also digital and an essential prerequisite to this study. The user can interactively drill down through multi-scale layers, with the cartogram algorithm operating at each layer. The algorithm is therefore applied recursively and in a nested hierarchical or tree structure (Cuff et al. 1984 present nested non-contiguous cartograms for representing groups of land use types within the boundaries of Pennsylvanian counties, though there is no tree structure used).

As yet cartograms have not been used for anything more than simple static maps, though Dorling et al. (2006) suggest the morphing of one cartogram into another in the context of an interactive interface (to emphasise unevenness in multiple globally-mapped social and economic variables – as opposed to unevenness of one variable across space).

However, cartograms have been criticised as being hard to understand or difficult to read (e.g. the perceived loss of geographic feature form and relationships noted by Gluhih and Portnov 2004). Shape is widely acknowledged as the variable to control this, but there is a trade-off between this and cartogram accuracy (Dent 1985, Tobler 2004). Many people also view cartograms as a novelty, a clever way of presenting information, but not as a map projection or as a tool to be used for serious analysis (Tobler 2004), though recent empirical evidence suggests that they are becoming increasingly popular (e.g. the WorldMapper project; Dorling et al. 2006). An objective of this research is to help reinforce this perception and present a useful application of cartograms that lies outside of its traditional use in the display of spatial data and information.

1.3 Objectives

The general purpose of this research is to develop a dynamic digital map that is more usable than traditional methods at communicating multiple scales. By implementing an automated cartogram algorithm (based on diffusion – Gastner and Newman 2004) to distort the shape of the map (though preserving the angles of intersection), an entirely new way of emphasising areas of interest or user focus is presented. The advantage is that the local view is presented whilst still preserving a global perspective. The objectives of this research are.

1. To develop a dynamic cartogram map that is based on the diffusion algorithm (created by Gastner and Newman) to distort polygons in a map to allow a spatially-enlarged point of interest.
2. To show that loss of shape is not a factor in spatial recognition.
3. To show that a cartogram can be used in this novel respect, to achieve a more intuitive local-global effect (as established through efficiency and satisfaction) than traditional global projection based maps using the zoom in/out method and the use of inset maps.

Therefore, a hypothesis can be stated that *use of cartograms is more usable (efficient, satisfying) at displaying multiple scales than the zoom in and inset map methods*. To establish whether this has occurred, tests were performed on the three methods to ascertain speed of navigation and correctness of orientation estimates (efficiency and effectiveness) followed up by a survey to gauge participant satisfaction (efficiency, effectiveness and satisfaction are the three variables conventionally used to ascertain usability – ISO 1998). The testing was backed up by a distorted shape and associated world map recognition exercise at the country / continent scale to isolate the effects of inherent shape recognition capabilities and associated world map geographical knowledge.

In summary, the use of a cartogram as a method of multiple scale display will be tested here alongside other established methods of managing representation of multiple scales. Since the cartogram demonstrates a fish-eye-like visual effect when used in this way, at least some part of any observed benefits of using cartograms could be down to the approximate fish eye distortion (though the fish eye method per se is not being tested here). Also, the specific (Gastner and Newman) cartogram algorithm itself is not being tested. A review is presented next, followed by an account of the methods used. The paper finishes with results, discussion and conclusions.

2 Review

2.1 The Cognitive View of Space and the GIS View of Space

The scientific view of space is that it is uniform and seamless – i.e. it exists as a continuous surface that extends infinitely in all directions. The cognitive view is not arranged in this way (Mark and Freundschuh 1994) and holds that geographical space can be organised into different categories that tend to be ordered into nested hierarchies (Peuquet 2002). This has been experimentally tested and confirmed by a number of researchers (Stevens and Coupe 1978, Eastman 1985, McNamara 1992). For example, people conventionally break down space into sub-spaces such as indoors and outdoors; these can then be further subdivided into smaller units.

We can carry these concepts over into the way in which we represent space within a computer. It has become standard practice to represent spatial entities as polygons, lines, points, which has cognitive agreement, though many entities are cognitively represented with indistinct boundaries (e.g. mountains, forests), which poses a challenge in the digital spatial domain (Mark 1999).

Spatial reasoning is another area where the structure and nature of space within a GIS does not match the way in which we experience and perceive space. Space is often formalised using a Cartesian co-ordinate system and described with vector algebra. This is different to the way in which humans draw conclusions about space. Accordingly, a qualitative method for the spatial reasoning of cardinal directions using an algebraic method has been presented, using the standard eight cardinal directions, (N, NE, E, SE, S,…) and a ninth symbol for directions of two points that were too close to each other (Frank 1995). We will carry this hierarchical and qualitative view of space through to the development and testing of the dynamic map.

2.2 Applications of Local Map Distortion

Another cognitively intuitive principle is the egocentric (as opposed to the common geocentric) view (e.g. Olson 1984; Gallistel 1990), emphasised by Hägerstrand with the presentation of his azimuthal projection (Hägerstrand 1957). This projection has the centre of enlargement at the point of interest (Figure 5.1). However, the function of this projection is to emphasise "real" distance (from the perspective of the individual, but also to model transport costs and their decrease with distance; Bunge 1966), while the approach of this paper is the enlargement of area (i.e. distance is not the focal spatial quality of the cartogram map).

The idea of fisheye views is analogous to that of placing a magnifying glass over the map. Furnas (1986) began applying the idea of fisheye views to not only maps but also to text in an attempt to solve the problem of presenting a global view whilst at the same time maintaining the local focus. He found that the fisheye view was far superior to other methods in allowing people to navigate file structures more quickly.

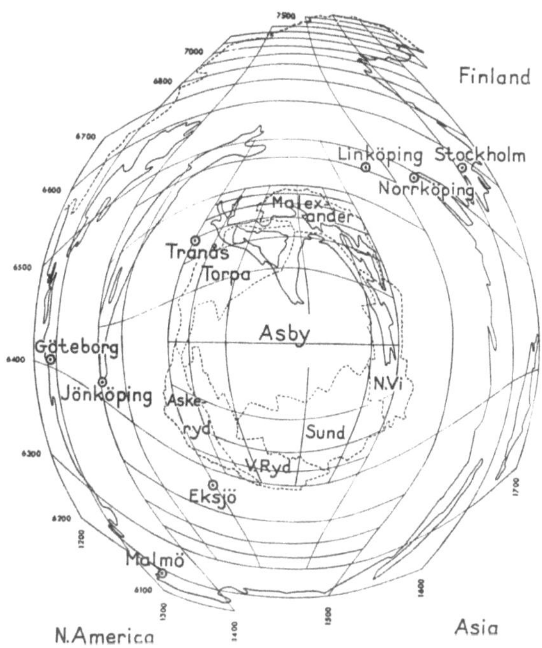

Fig. 5.1. Hagerstrand's egocentric azimuthal (logarithmic distance scale) projection of central Sweden (Hägerstrand 1957)

Sarkar and Brown (1994) state that the advantage of using a fisheye view over that of the zoom in / zoom out method conventionally used in GIS is that it allows the user to view the local area of interest, whilst still maintaining a reference to the global view (Sarkar and Brown 1994, p 73). Gutwin and Fedak (2004) noted that people carried out web navigation tasks significantly faster with the fisheye view than a two-level zoom. However, Carpendale et al. (2004) note limitations at high magnification and suggest alternative distortion functions to enable this, including multiplicative magnification produced by superimposing one lens on top of another. Other examples of local distortion include using projections to distort specific areas (i.e. can also be polyfocal – Kadman and Shlomi 1978; Yang et al 2000).

There are also precedent projective examples for accentuating areas of user interest. Snyder (1987) presented a dual projection approach, simulating a magnifying glass centred on an area of interest within a globally-projected map (i.e. a magnified circular or rectangular area surrounded by a conventionally-projected background). Leung (1989) applied a similar bifocal principle to London Underground maps. Harrie et al. (2002) adopt a continuously varying approach, similar in transformation terms to Hägerstrand's projection. Finally, Cuff et al. (1984) adjust cartogram areas (originally calculated for land use) based on reader perception, which has an affinity with the approach adopted in this paper.

2.3 Automated Cartograms

Since Tobler's original automated cartogram in the 1960s, which strived to preserve shape as much as possible (Tobler 2004; Gastner and Newman 2004), many variants have been developed. These include, use of a hexagonal grid to effect cartogram distortion (reported on in Tobler, 2004), rubber sheeting (Ruston 1971), implementation of push-pull forces from polygon centroids to manipulate boundary coordinates, and therefore area (Dougenik et al. 1985), application of fine-resolution cellular automata to redistribute area (Dorling 1996), vertex shifting whilst retaining position of "important" points to maintain an essential shape (Kocmoud and House 1998), use of polygon medial axis lines to structure area (Keim et al 2005), use of Delaunay triangulation (Inoue and Shimizu 2006) and algorithms to construct rectangular cartograms (van Kreveld and Speckmann 2007). It should be noted that these are all areal as opposed to linear cartograms.

Gastner and Newman (2004) developed an algorithm that calculates contiguous cartograms based on the linear diffusion process of elementary physics. The general idea is to take a standard map and allow the areas of high density of some attribute to flow to those of low density until a point of equilibrium is reached.

The attribute is initially described by a density function $p(r)$, where r is used to represent the geographic position. The premise is that over time the density will diffuse until it becomes uniform and so come to rest. The total displacement of r from start to finish represents the new projection needed to produce the density equalising cartogram (Figure 5.2). The method can be likened to Dorling's cellular automata algorithm in that different regions trade area until a fair distribution is reached.

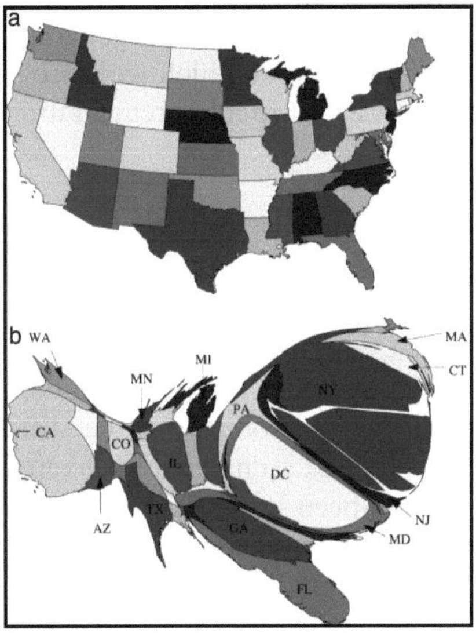

Fig. 5.2. Distribution of news stories by state in the United States. (*a*) a true area depiction (Albers conic projection). (*b*) Cartogram (sizes of states are proportional to their frequency of mention in news stories). (Gastner and Newman 2004)

Gastner and Newman's algorithm will be used for this project as it exhibits a good trade off between shape recognition and accuracy (this is an attribute it shares with a few other methods e.g. Dorling 1996; Kocmoud and House 1998; Keim et al. 2005). It also performs areal shape distortion locally whilst maintaining the form of peripheral areas as much as possible. Practically, the source code was freely available for use in this project (Gastner 2005).

It is important to note that the cartogram solution is conformal and that any apparent spatial distortion is of areal shape only. Therefore when distortion is referred to in this paper it is in the areal shape context only.

2.4 Summary

The way in which human beings construct the world around them in their minds is a complex process and is difficult to reproduce within the digital domain. Cartograms perhaps offer us a way that can match the local and global spatial distortion we sub-consciously construct in our minds. The cognitive view of space is also a discretely categorised and hierarchical one. Most of the cartogram techniques presented, while distorting space continuously, do maintain some semblance of the discrete boundaries of areal entities that occupy that space. Therefore, in adopting cartograms we can perhaps create a framework for viewing geographical information that is much more attuned to the way in which we store geographical space in our mind, and therefore are easier to interpret and use.

3. Methods

3.1 The Participants in the Study

Thirty participants were selected and tested in order to ensure sufficient numbers to run statistical tests comparing the different methods. Most were either in university or had a university background and were known to the lead author. Having been informed of the experimental details, all participants consented to the experiment. They were all over the age of eighteen and a wide demographic of people was surveyed. This ranged from people in their teens and early twenties who had some or a lot of experience with computers to older people in their forties and fifties with limited experience of using digital maps and computers.

3.2 Resources Used in the Study

3.2.1 Data Preparation

Prior to development of the cartogram map, the spatial data that it displays needed to be prepared at each level of detail. All data was drawn from the LINZ New Zealand topographic data set. At the top level (representing the smallest scale) is the outline of New Zealand and constituent provinces (Figure 5.3 shows the South Island of New Zealand at this scale). Each province was composed of several smaller polygons representing districts within the province (second level), the districts were in turn composed of different city, town and rural areas (third level), which were themselves composed of different suburbs in the case of cities (fourth level). The display of one of these suburbs constituted the fifth and finest level of detail. Roads were displayed at this scale (above 1:250,000; e.g. 1:50,000) as there are no polygon shapes corresponding to this level, yet visual cues are needed (Figure 5.4 shows the cartogram map at each level). This data was fed into the cartogram generation source code (Gastner 2005) as needed, based on user choice regarding the map.

3.2.2 The Cartogram Map Application

This section outlines the development of the map and the different components that constitute the major structure of the program. These are specifically the tree structure used to create the hierarchy and store the data, the cartogram engine and the map display itself.

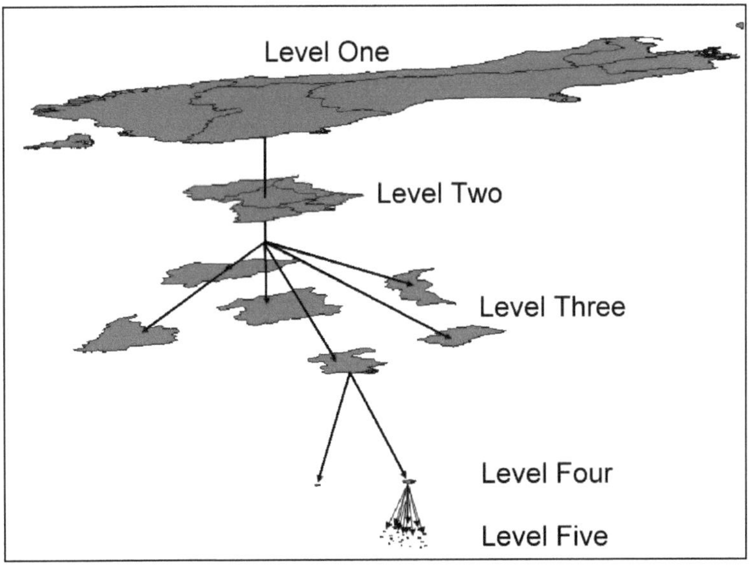

Fig. 5.3. The hierarchical structure of the cartogram map data

The role of the cartogram generation algorithm (Gastner and Newman 2004) was to distort polygon shapes within the application according to user focus or attention, to effect a scale change to "magnify" the area of focus. Some modifications were made to the supplied source code for the purposes of the cartogram map application. These were mostly effected to make the algorithm faster. For example, given the vague (hard to quantify) nature of user focus, procedures that optimised the accuracy of the cartogram areas were suppressed, importantly also promoting retention of the original shape as much as possible. Another modification saw the constraints against using lines for distortion being removed. This allowed the algorithm to then distort road features at the fifth level of detail.

3.2.3 Software and Hardware

The cartogram map interface was written in Java (a front end for the underlying C algorithm code), well known for its enhanced graphical user interface capabilities. It consisted of a main viewing frame that showed the map data and a series of buttons across the top which activated controls such as zoom in/out, inset map and activate cartogram (Figure 5.4).

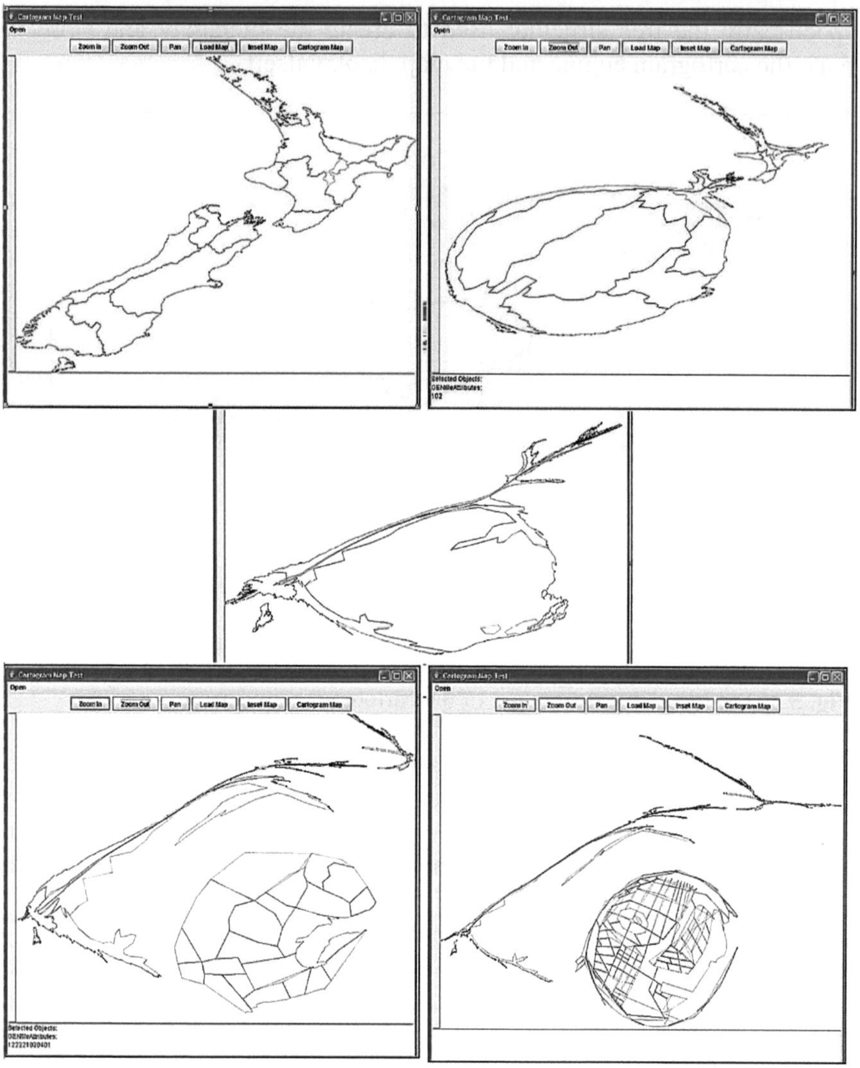

Fig. 5.4. Screen shot of the cartogram map at each level for Dunedin: a) New Zealand; b) Otago province; c) Dunedin city limits; d) Dunedin suburbs; e) city centre suburb with roads

When a user activated the cartogram map, any button click within a polygon on the displayed map (initially set at the top level) expanded that polygon using the diffusion algorithm, displaying the spatial data at the next finest level of detail for that polygon, creating a fisheye-like effect overall. What has really happened is that the coordinates of the sub-polygons replace the coordinates of the original clicked polygon that covers them. This amended dataset (with all other unclicked polygons remaining the same) is input into the algorithm, along with a relative weighting that reflects user focus. For ease of display, the clicked area was assigned a value of 500 while the unclicked areas were assigned a value of 1. These weightings led the cartogram area assignation for the ensuing display. Finally, all tests (to be described in detail in the next couple of sections) were carried out on the same Sony Vaio laptop with a 15 inch screen and a 1.2 GHz processor.

3.3 Experiment Design

The thirty participants were broken down into five groups of six participants. Each participant took tests associated with each of the three methods, zoom in, inset map and the cartogram. Each method would involve navigating to one of three different locations. The order in which a given method was used to navigate was selected randomly as was the location. This was done to ensure that there was a balance in any learning carry over effect from the previous methods used. Each group of six would ensure that each combination of method and location was used. Comparisons were made within subjects (using ANOVA tests that ignored the groupings) and not between subjects.

3.4 The Procedure

The experiment was in three parts. Before the testing of the cartogram map described in the last section, a test of country and continent shape recognition was made. This is to indicate basis of performance in the subsequent testing of the maps. Would the participant be reacting to inherent distortion due to the algorithm rather than any lack of basic shape recognition skills? Accordingly, a series of five different commonly known countries and continents (Australia, the British Isles, South America, New Zealand and the United States of America) that had had their shapes either expanded or shrunk by the algorithm, were shown to the participants (see Figure 5.5 for examples). Participants were asked if they recognised the distorted country or continent and if so what they believed it to be. At this stage, no change of scale was attempted or tested and therefore no comparison with zooming or inset maps made.

The main part of the survey involved testing the cartogram map and the other techniques for finding local views. The hypothesis to be tested has been expressed in section 1.3. The cartogram-enabled dynamic map was tested alongside two other methods of changing map scale: the standard zoom in/out and a modified zoom in/out with an inset map. Three locations were chosen as the focus of zooming: Queen Street in Auckland, Seymour Square in Blenheim and the Octagon in Dunedin. Each participant zoomed in to one of three locations using one of the three tested methods.

Each participant was asked to navigate to the three different locations within New Zealand. For each location the participants used one of the three tested methods using the full monitor display. The participants were timed as they explored their particular location-method combination (see figure 5.4 for the Dunedin set of maps), giving an indication of efficiency of the tested technique and implicitly, how usable and perhaps, how cognitively intuitive it is. Computer processing time was not considered a component of the measured time.

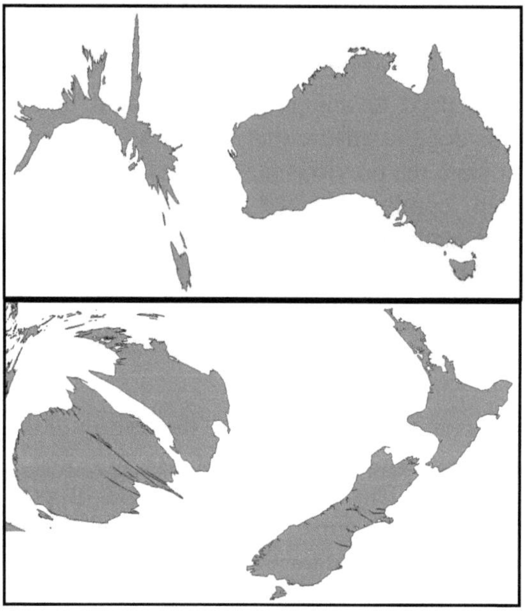

Fig. 5.5. Comparison of country shape with their cartogram equivalents. a) a shrunken Australia; b) an expanded New Zealand

Once the participants had successfully navigated to their specified location they were then asked three questions relating to the relative orientation of other New Zealand cities and towns. The orientation was specified as a choice of general cardinal directions (north, north-east, east, south-east, south, south-west, west and north-west), as proposed by Frank (1995). The purpose of the orientation questions was to test whether having a global perspective present allows people to have a better sense of their location within space compared to those with just a local view.

The third part of the survey asked participants general questions about what experience they had had with digital maps, their personal knowledge of geography and importantly, their satisfaction with their experience with the cartogram map. This indication of satisfaction is also an important indication of how useful and usable the participants found the interface. A comparison was made of how the participants performed in the map tests with their experience with maps and geographical knowledge.

The basis of the fourth and final part of the survey was a world map. The participants were given a list of the same five countries and continents as tested for distortion in the first part. The participants then had to mark where each of the countries or continents was on the world map. The purpose of this was to ensure that any incorrect answers in the first part of the survey were not a result of lack of geographical knowledge and indeed from non-recognition. This was placed last and was kept separate from the survey sheet handed to the participants so that they could not use it as a reference for the other parts of the test.

All results were entered and stored in a spreadsheet so that the information could then be sorted much more easily and made ready for statistical analysis. This then allowed for a comparison between each method to see if there was any statistically significant difference in the time taken.

4. Results and Analysis

4.1 Distortion of Shape

In Part A of the survey participants were asked if they recognised the distorted country or continent shown and if so asked to write what they believed it to be. The answers were either correct or incorrect and a count of correctly identified countries and the percentage for each is shown in Table 5.1.

Table 5.1. Correctly identified countries

Country	Correct	Incorrect	%
Australia	16	14	53%
British Isles	20	10	67%
New Zealand	21	9	70%
South America	20	10	67%
USA	23	7	77%

The way in which the country was distorted was random. The following distortions were used: Australia and the United Kingdom had been shrunk whilst New Zealand, USA and South America had been expanded. The results displayed are as expected, given the characteristics of the participant population, and indicate good recognition. Wiegand and Stiell (1996) reported on recognition scores for undistorted continents (the top three recognised were Australasia - 72%, Europe - 57% and Asia – 51%), but for English Year 6 children.

For this study, recognition was largely uniform except for the Australia figure (5.5a), with the least amount of participants (53%) identifying it correctly. The results were supported by those from the last stage of testing. The participants' geographical location knowledge of the shapes tested was 100%, so any variation in the shape recognition results can be said to be down to shape recognition skills.

Informal feedback revealed that participants recognised the shapes on the basis of specific features (e.g. Cape Horn on South America). It is the considerable distortion of features existing at this level of recognition that seems to have stymied widespread recognition of Australia (e.g. the elongated Gulf of Carpentaria). Removal of peripheral land masses to isolate the shape being tested (e.g. the European mainland in the case of the British Isles) may also have confused participants, as did the algorithmic tendency for compression along the y-axis (which made South America appear like Africa for some participants). Factors such as these and the fact that shrinking was applied (the best recognised countries USA and NZ – figure 5.5b – were both expanded) may account for the misrecognition of Australia.

4.2 Comparison of Different Zooming Techniques

The second part of the survey was composed of two tasks. The first task was to navigate to one of the three New Zealand locations. The time taken to reach the location was recorded to be used as the raw data. The second part of this stage was to indicate which direction they believed other locations were in relation to the point they had navigated to. The box and whisker plot (Figure 5.6) shows the time taken (in seconds) by each participant to navigate to the required location for each method. The summary statistics are presented in table 5.2. The results show that the fastest method is method three, the cartogram map, the second fastest is the inset map method and the slowest method is traditional zoom in/out method.

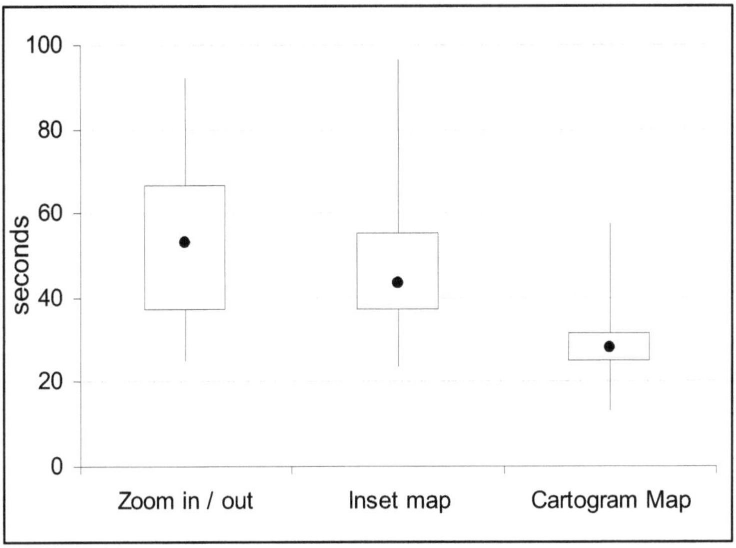

Fig. 5.6. Box plot of time taken for each method

Table 5.2. Summary statistics for time taken for each method

	Zoom in/out	Inset map	Cartogram Map
minimum	24.90	23.50	13.00
lower quartile	36.83	37.03	24.63
median	53.20	43.40	28.20
upper quartile	66.88	55.30	31.40
maximum	92.10	96.70	57.40
average	54.15	45.77	30.28
standard deviation	19.12	14.17	9.97

The results for the different methods of multi scale representation were initially assessed for significance using a within-subject ANOVA test. The F ratio derived and the associated p-value being less than the 0.05 threshold [$F(2, 87) = 16.27$; $p<0.0001$] indicates that there is a significant difference between groups of values (F ratio is larger than the null hypothesis value of 1).

4.3 Orientation with Different Methods

When each participant had successfully navigated to the required location, they were then asked to state which direction they believed three different locations to be in. The choices they were given were the eight cardinal directions. A summary of the number of correct responses for each method is given in Table 5.3.

How the participants responded according to the four main cardinal directions - North, South, East and West - was then calculated. The results for these, which show a marked improvement in the number of correct responses over eight cardinal directions, are shown in Table 5.4. ANOVA tests for all orientation results proved not significantly different from chance.

Table 5.3. Count of correct responses (out of eight cardinal directions) for each method

Number correct	Zoom in/out	Inset Map	Cartogram Map
0	3	3	6
1	9	13	10
2	17	10	6
3	1	4	8
Average	1.53	1.5	1.53

Table 5.4. Count of correct response for four cardinal directions

Number correct	Zoom in/out	Inset Map	Cartogram Map
0	0	0	0
1	0	0	0
2	6	5	1
3	24	25	29
Average	2.80	2.83	2.97

4.4 Participants' Perceptions

In the third section of the survey participants were asked seven questions, five indicating degree of satisfaction with the cartogram map, and two relating to experience with New Zealand geography and digital maps. Participants were given a range of responses from one to five, one meaning that they strongly agreed with the question and five if they disagreed. The question and modal values are shown below in the summary Table 5.5.

In order to test whether the results from the 5 point scale occurred because of the pooled preferences of the participants and are significantly different from a random sample of values (i.e. so that we can say that, for example, the strong modal preference for question 2 is actually statistically valid), a chi squared test was performed. From the results shown in Table 5.5, all but the question regarding New Zealand geography are statistically significant.

Table 5.5. Summary Statistics for Participants Questions

Question	Mean	Mode	Std	Chi
Structure matches way I organise space in my head	1.97	2	1.03	0.0001*
Shape is important to me in recognition	1.53	1	0.94	0.0001*
I still recognise shape after the transformation	2.57	3	1.10	0.0120*
Having the global perspective helped	1.77	1	0.97	0.0000*
Structure easier to use	2.17	2	1.05	0.0067*
Personal knowledge of NZ geography	2.53	2	1.20	0.45300
Experience with digital maps	3.00	3	1.44	0.0113*

5 Discussion

This section will discuss the results and analysis. It will compare the results obtained from the experiment with participants to objectives of the research. It will also try to explain the possible reasons why the results occurred in the way they did.

5.1 Distortion of Shape

The first section of the survey was intended to ascertain whether shape was an important feature in recognition. The results showed that different countries and continents had different levels of recognition, though the vast majority were recognised, therefore the assertion can be accepted.

The results of the distortion of the shape and the recognition can be related to the way in which the algorithm works. Generally the centre of the shape remains in the same place; however the vertices around the shape are displaced in order to increase the area. This means that a shape that has a greater density of points along its edge is likely to be distorted more than a shape with a smaller density of points. As a result of this, those areas that are distinctive, such as Cape Horn in South America, can become more distorted as they generally contain more vertices and therefore are subject to greater treatment by the algorithm.

5.2 Comparison of Map Methods

The aim of this section of the survey was to establish whether the cartogram map is more efficient than other methods of zooming to areas of user interest. If we take the measure of time as a coarse measure of how usable a method is, then we can state the cartogram map is the most usable method followed by the inset method then the traditional zoom in and out method. Although some participants had expressed difficulty in deciphering what the distorted map was representing, the results show that this was not a major obstacle in the use of the map. Therefore, the hypothesis expressed in section 1.3 can be accepted.

The results also indicate the extra work required by the weaker-performing methods. The difference between those methods can be attributed to the use of the inset map as a guide. The inset lets the user have a visual cue as to where they currently are in the global view, unlike the traditional zoom in and out method which leaves the user blind in this respect.

5.3 Comparison of Orientation Results with Different Methods

The most obvious explanation for the non-significant (consistent) results for this sub-stage of the test is that participants relied on their knowledge of New Zealand geography and this was sufficient to produce a result that matched the other methods overall. This negated the relative qualities of the three methods and could explain why the average score for each method effectively matches the average of the other methods.

Another factor that could have caused the third method to perform less well than expected is that the global view it presented was wildly distorted, shifting locations from their expected place on the map to somewhere else with a different relative orientation. Therefore most uses of the global view as a true account of the relative position of a location are incorrect.

5.4 Discussion of Participants' Perceptions

The final set part of the experiment asked the participants a series of questions about how they perceived space and interacted with the cartogram map as indications of satisfaction with the map, another facet of cognitive affinity. The results for the five questions that relate to this were significantly different from random, indicating true user preference.

The first question asked the participants to what extent they believed the hierarchical structure of the map matched the way that they organised space in their own heads. Their modal answer that the structure was a good match to how people cognitively categorise space within their mind was reinforced by the efficiency results. These two results combined would suggest that the assertion that the cartogram map would prove to be more usable is supported.

The third question in the survey asked participants if they still recognised the shape of the cartogram map after the transformation, an important question given the strong emphasis that people give shape in recognising objects. The neutral modal answer suggested that many people did not recognise the features in front of them once the transformation had occurred, which was surprising since the transformation is triggered by the user and given the efficiency results. However, due to the modifications made to the algorithm code to preserve shape at the expense of accuracy, the cartogram appears to have become non-contiguous at the more detailed levels (at levels 4 and 5 in the case of Dunedin – Figure 5.4d and e). Whether this had an adverse effect on testing was not highlighted.

The fourth question asked the participants whether they believed having the global perspective shown at the same time as the local view helped them orientate themselves better. They believed that the use of a global perspective was a very important factor in aiding them to better orientate themselves when trying to decide where they are placed within space.

This result is not matched with results of orientation tests, despite some encouraging results with the 4-direction tests (zoom vs. dynamic). Given the high results for the personal NZ geography question, though not significant, the poor performance with orientation can be attributed to the cartogram algorithm.

The fifth question asked participants if they found the structure of the cartogram map easier to use than the traditional zoom in and out method and inset methods. The modal answer here was two; giving support to previous results that showed the map was more usable.

6. Conclusions and Future Work

6.1 Conclusions

A dynamic map has been developed that uses an automated hierarchical cartogram to effectively replicate the action of a zoom tool. Like other egocentric maps (increasingly used in personalised and mobile spatial data contexts), the local map view has been emphasised whilst still retaining the global context.

The construction of the cartogram map (using Gastner and Newman's diffusion algorithm) fulfills the first objective, as stated in section 1.3. Overall, it was concluded that loss of shape is not a factor in recognition (second objective), though there is more rigorous testing needed to more firmly establish this, especially in the case of shrunken shapes. This justified the use of shape-distorting cartograms (at least in an expansion context), the mechanism behind the cartogram map. The third objective was achieved and the stated hypothesis was accepted, in that the presented method was demonstrated to be more usable as a multiple scale representation (in the areas of efficiency and satisfaction) than established methods such as zoom in / out and inset maps. However, results for the orientation aspect of this objective proved inconclusive.

Aspects of the collected dataset remain unexplored. Although the cartogram map performance has been established in both the areas of efficiency and satisfaction alone, no effort has been made to link the two parameters together. This approach would mean an analysis of whether the participants' opinions actually translate into actions in accordance.

The following should be emphasised as innovative aspects of this research:

1. The use of the cartogram as a means to replicate the zooming function is in itself novel. Cartograms have inherent features that facilitate this role:
   a. They specifically target geographical features for distortion and has inbuilt functionality for the optimised retention of shape. This makes it especially valuable as a potential technique to communicate both local and global geography without recourse to its geographically correct counterpart. Local projective and fisheye maps do not address geographic feature shape directly, increasing the likelihood that the map will be distorted beyond recognition and negating their value for zooming.
   b. Another desirable aspect of using cartograms is that distortion is a truly local process. Areas not affected by user focus or interest undergo minimal change. This is not the case with projective solutions – although the emphasis in this case is on the local enlargement of space, the transformation is performed globally.
2. The implementation of the cartogram in this context has engendered a hierarchical or nested structure, with the cartogram algorithm operating at each level. Therefore it operates from a recursive tree structure.
3. It is an interactive cartogram in that the user can change or refine user focus by engaging with a software interface.
4. The assigning of user focus as a weighting is a novel variable for the distortion of area central to the cartogram's operation. The use of such a vague variable for cartogram area calculation frees us somewhat from some of the constraining aims of cartograms.

a. Firstly, the somewhat arbitrary nature of the user focus weighting means that the area of the distorted region only has to be approximately true to the weighting.
b. Furthermore, the aim to retain original shape as much as possible can be relaxed as distortion is essential to achieving the local-global map (though enough semblance of the original shape still has to be retained).

6.2 Future Work

An obvious next step is the testing of the dynamic cartogram map alongside egocentric maps enabled by local projections or fisheye views. The testing should encompass an explicit cognitive test (e.g. cognitive walkthrough) to establish effectiveness as well as the already tested aspects of efficiency (time taken) and satisfaction (as indicated by the participant survey).

Other interesting facets of the cartogram map that could be explored in the context of this proposed test are the limitations of the hierarchical structure, which forces the user along preset pathways of navigating multiple scales which may be at odds with their desired path, unlike the projective methods which allow unconstrained exploration. This could be part of an effort to isolate the contribution that the hierarchy makes (as opposed to the cartogram itself) to the observed performance of the cartogram map.

The unresolved question of orientation can be investigated, whether it is a function of the distance of the location from the centre point, or whether it is a result of how far north, south, east and west the point is. This testing could be based on explicitly using the four cardinal directions, rather than deriving them from the results as was done above.

Another area that could be investigated further is the nature of recognition with regard to different distortions of shape. The results of the survey suggested that different distortions caused different levels of recognition. A more thorough investigation into whether shrinking a shape makes it harder to recognise than an enlarged shape could be devised, to investigate the observed bias towards recognising expanded shapes. The same shape could be both expanded and shrunk to ascertain the overall effect.

The weighting assigned as a user interest variable to the cartogram algorithm is important as it is an explicit driver of area and therefore shape. However, the weighting level was arbitrarily set, at 500. To address this, the map could be easily modified to let the users set their own level of interest in the map. The map could then be tested again to see how it performs when the user is given more control over the environment that they are exploring. Indeed, user performance can be tested at different weighting levels in order to identify an optimal level of weighting for shape recognition.

Nevertheless, there is always the question to address of whether the cartogram should always be unaided in case recognition is difficult or impossible (which is a challenge for this technique and indeed any other cartogram). Knowledge of the geography being represented as a transformed map will always be a factor for the individual's recognition of map elements (it could be argued that the experiment never tested this through using base maps that were likely to be well known to any participant). It could also be argued though that an undistorted map of an unfamiliar area could also foster non-recognition, which makes that map as ineffective as a confounding cartogram.

However, by definition cartograms do have the capability of transforming familiar geographic features into unrecognisable entities and for that reason aid should be available for map readers that experience this. A simple solution could be to show the unprocessed map alongside the cartogram (e.g. Figure 5.5). However, a good compromise to preserve elegance of display would be to use a roving tool that reversed the effects of cartogram distortion to selectively reveal the original map. This reverse fish-eye-like view would have broad applicability for cartograms, and could be especially useful in the public understanding of them.

Acknowledgements

The authors would like to thank Brian Niven for his statistical help in the experiment design. Thanks also to the reviewers for their helpful comments and references, especially the one who suggested the reverse fish-eye-like view on a cartogram. Finally, we are also grateful to the survey participants for their invaluable role in the study.

References

Bunge W (1966) Theoretical Geography, 2nd edn. Gleerup Publishers, Lund, Sweden

Carpendale S, Light J, Pattison E (2004) Achieving Higher Magnification in Context. Proceedings of the 17th Annual ACM Symposium on User Interface Software and Technology, ACM Press, New York, USA, pp 71–80

Cuff DJ, Pawling JW, Blair ET (1984) Nested value-by-area cartograms for symbolizing land use and other proportions. Cartographica 21(4):1-8.

Dent B (1985) Principles of Thematic Map Design. Addison-Wesley, Reading, Massachusetts

Dorling D (1996) Area cartograms: Their use and creation. Concepts and Techniques in Modern Geography (CATMOG), No 59

Dorling D, Barford A, Newman M (2006) WORLDMAPPER: the world as you've never seen it before. IEEE Transactions on Visualization and Computer Graphics 12(5):757-764

Dougenik JA, Chrisman NR, Niemeyer DR (1985) An algorithm to construct continuous area cartograms. Professional Geographer 37(1):75-81

Eastman JR (1985) Graphic organization and memory structures for map learning. Cartographica 22(1):1-20

Frank AU (1995) Qualitative Spatial Reasoning about Cardinal Directions. In: D Mark, D White (eds) Proc. Autocarto 10. Baltimore, MD, pp 148-167

Furnas G (1986) Generalized Fisheye Views. Proceedings ACM CHI'86, pp 16-23

Gallistel CR (1990) The Organization of Learning. MIT Press, Cambridge, USA

Gastner M, Newman M (2004) Diffusion-based method for producing density-equalizing maps. Proceedings of the National Academy of the Sciences (PNAS) 101(20):7499-7504

Gastner M (2005) Michael Gastner homepage, http://www.santafe.edu/~mgastner (accessed 6th May, 2008)

Gluhih R, Portnov BA (2004) Visualization of the spatial patterns of inter-urban income disparities using coordinate transformations. International Journal of Geographical Information Science 18(3):281-297

Gutwin C, Fedak C (2004) Interacting with big interfaces on small screens: a comparison of fisheye, zoom, and panning techniques. Proceedings of the 2004 conference on graphics interface. ACM Press, New York, USA, pp 145–152

Hägerstrand T (1957) Migration and area. In: Migration in Sweden, Lund studies in Geography, Series B, Human Geography, No.13. Department of Geography, University of Lund, Lund, Sweden, pp 27–158

Harrie L, Sarjakoski LT, Lehto L (2002) A variable-scale map for small-display cartography. Proceedings of the Joint International Symposium on "GeoSpatial Theory, Processing and Applications" (ISPRS / Commission IV, SDH2002). Ottawa, Canada, July 8-12, 2002, CD-ROM

Inoue R, Shimizu E (2006) A New Algorithm for Continuous Area Cartogram Construction with Triangulation of Regions and Restriction on Bearing Changes of Edges. Cartography and Geographic Information Science 33(2):115-125

International Standards Association (ISO) (1998) Guidance on Usability. In: ISO 9241-11 Ergonomic requirements for office work with visual display terminals (VDTs) – Part 11

Kadman N, Shlomi E (1978) A polyfocal projection for statistical surfaces. The Cartographic Journal 15(1):36 – 41

Kiem DA, Panse C, North SC (2005) Medial-axis-based cartograms. IEEE Computer Graphics and Applications 25(3):60-80

Kocmoud CJ, House DH (1998) Continuous Cartogram Construction. Proc. IEEE Visualization, pp 197-204

Leung YK (1989) Human-computer interface techniques for map based diagrams. In: G Salvendy, MJ Smith (eds) Proceedings of the 3rd International Conference on Human-Computer interaction on Designing and Using Human-Computer interfaces and Knowledge Based Systems, 2nd edn. Elsevier Science, New York, NY, pp 361-368

McNamara TP (1992) Spatial Representation. Geoforum 23(2):139-150

Mark DM, Freundschuh S (1994) Spatial Concepts and Cognitive Models for Geographic Information Use. In: TL Nyerges, DM Mark, R Laurini, MJ Egenhofer (eds) Cognitive Aspects of Human Computer Interaction for Geographic Information Systems. Kluwer, London, pp 21-28

Mark DM (1999) Spatial Representation: A Cognitive View. In: P Longley, M Goodchild, D Maguire, D Rhind (eds) Geographical Information Systems vol 1. John Wiley and Sons, New York, pp 81-99

Meng L (2005) Egocentric Design of Map-Based Mobile Services. The Cartographic Journal 42(1):5-13

Olson J (1984) Cognitive issues in map use. International Yearbook of Cartography, pp 151-157

Peuquet DJ (2002) Representations of Space and Time. The Guilford Press, New York

Ruston G (1971) Map transformations of point patterns: Central place patterns in areas of variable population density. Papers and Proceedings of the Regional Science Association 28:111–29

Sarkar M, Brown M (1994) Graphical fisheye views. Communication, Association of Computer Machinery 37(12):73-84

Snyder JP (1987) "Magnifying-Glass" azimuthal map projections. The American Cartographer 14(1):61-68

Stevens A, Coupe P (1978) Distortions in judged spatial relations. Cognitive Psychology 10:422-437

Tversky B (1992) Distortions in cognitive maps. Geoforum 23(2):131-138

Tobler W (2004) Thirty Five Years of Computer Cartograms. Annals of the Association of American Geographers 94(1):58-73

Van Kreveld M, Speckmann B (2007) On rectangular cartograms. Computational Geometry 37:175-187

Wiegand P, Stiell B (1996) Lost continents? Children's understanding of the location and orientation of the Earth's masses. Educational Studies 22(3):381-392

Yang Q, Snyder J, Tobler W (2000) Map projection transformation: Principles and applications. Taylor & Francis, London

CHAPTER 6

Exploring Tessellation Metaphors in the Display of Geographical Uncertainty

Julian Kardos[1], Antoni Moore[2] and George Benwell[3]

[1]Intergraph Corporation New Zealand, Auckland, New Zealand

[2]School of Surveying / Department of Information Science, University of Otago, Dunedin, New Zealand

[3]School of Business, University of Otago, Dunedin, New Zealand

Abstract

This paper explores in detail a new and effective metaphor for visualising choropleth map uncertainty. The "level-of-detail" metaphor has been shown here to communicate attribute uncertainty, but also spatial uncertainty as a secondary expression. The metaphor is delivered to the map viewer via the regular tessellated output of the Hexagonal or Rhombus (HoR) quadtree spatial data structure, as a semi-transparent map layer that lies on top of the choropleth (termed the *trustree* when used in this manner). For testing, multiple images were created with differing resolution levels of output from the *trustree* and superimposed on a New Zealand 2001 census choropleth map of Dunedin City. An Internet survey was designed and run, to reveal the visual metaphors that the *trustree* communicates uncertainty through. The choice of metaphor offered was (1) a level of detail (or resolution) metaphor, where less detail (i.e. coarser resolution cells) represents more uncertainty (i.e. uncertainty is sketchy), or (2) a metaphor of clutter, where the data structure output can be sufficiently dense so as to cover spatial information, in effect hiding uncertain areas (i.e. uncertainty is a barrier). In this case the finer resolution cells indicate more uncertainty. Also, the survey aimed to determine a *usable trustree* tessellation resolution level to express uncertainty information. The results showed the *trustree* tessellation was more effective when representing a metaphor of detail and that attribute and spatial uncertainty can be effectively expressed, depending on the tessellation level used.

Keywords: choropleth, spatial uncertainty, attribute uncertainty, image schema, quadtree, detail, clutter

1 Introduction

New and intuitive methods for the representation of spatial and attribute uncertainty can increase the validity of and speed to arrive at decisions using spatial data. Increasingly, spatial data are used in analysis and modelling contexts in environmental and socioeconomic applications (Heuvelink and Burrough 2002). These are vital real world contexts, so the scope for uncertainty representation methods is broad and sorely needed to decrease the likelihood that time and money are lost in pursuing a false decision avenue. The question of how uncertainty visualisations work in such complex decision making scenarios is a current research question addressed and reported on in MacEachren et al. (2005) and Hope & Hunter (2007a, 2007b).

However, the focus of this paper is purely on the representation of spatial and attribute uncertainty and how that representation works on a metaphorical level. Expressing uncertainty in spatial data is a key research challenge on the geovisualisation agenda put forward by MacEachren and Kraak (2001) and refined by Fairbairn et al. (2001) from the point of view of representation. Updates to this agenda as a whole are addressed in Kraak and MacEachren (2005) and the Dykes et al. (2005a) book, specifically Dykes et al. (2005b). More recently, MacEachren et al. (2005) assert that "…[u]ncertainty in geospatial information is a fundamental issue … [but] we have only scratched the surface of the problem." Addressing this research challenge, a new technique for visualising uncertainty was developed, the *trustree* - '**T**he **R**epresentation of **U**ncertainty using **S**cale-unspecific

Tessellations in **T**ree form' (see Kardos et al. 2005 for more detail on the *trustree* method, and Kardos et al. 2007 for exploration into choropleth mapping techniques and map legends). The Hexagonal or Rhombus (HoR) quadtree (Bell et al. 1989; referred to in the current context as the *trustree*, to distinguish this from the conventional usage of the quadtree data structure as a means of efficient storage) was used to express attribute and spatial boundary uncertainty. The output from the quadtree, a regular hierarchical tessellation, was overlain on a choropleth map of census data (Figure 6.1). The size of the quadtree cells communicated the amount of uncertainty possessed by the census data displayed beneath, which is uniform for a census polygon. In the superimposing *trustree*, larger cells expressed greater uncertainty (section 3.3 specifies the quantitative relationship between cell size and uncertainty used).

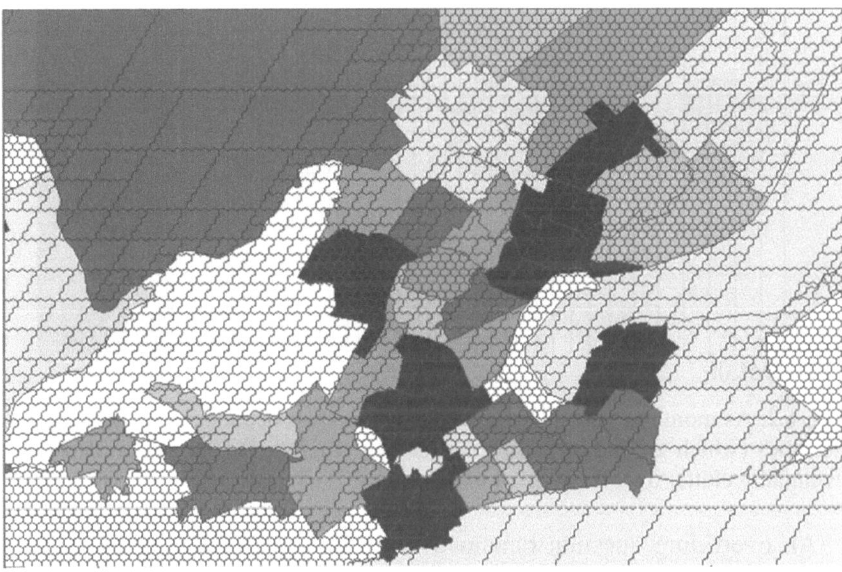

Fig. 6.1. A screenshot of the original *trustree*. The larger the cell in the tessellation overlying the choropleth map, the more uncertainty is implied

The *trustree* also distributes a linear choropleth spatial boundary (in this instance, that of a census tract) into areal tessellations. Therefore, the uniform uncertainty possessed by a choropleth polygon has been visually transformed into a semi-continuous form that better reflects the "truth" of uncertainty distribution. Although the size of quadtree cell is calculated from attribute uncertainty, it actually communicates spatial uncertainty associated with choropleth boundaries, which are for the most part geographically arbitrary. Spatial uncertainty is not expressed in a quantitative sense, the transition of cells across a boundary de-emphasises that boundary; therefore uncertainty in qualitative form is conveyed to the map viewer.

The ability to represent more than one type of data uncertainty addresses a challenge for uncertainty geovisualisation presented by MacEachren et al. (2005): "…little has been done to address the challenge of depicting multiple forms of uncertainty in the same display." This paper will explore this capability further by looking at how the metaphors used by the *trustree* (or rather their underlying image schema) allow such multiplicity to happen.

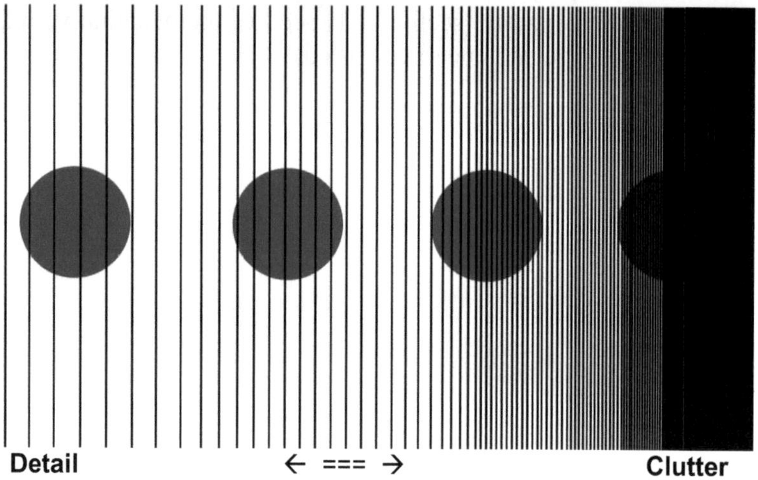

Fig. 6.2. As more lines are added over the circles their detail becomes lost because the lines (which are indicative of *trustree* cell resolution) become too dense and eventually clutter over their visual impression

An overriding question remained from the research described above – *does the trustree visualisation work more effectively when smaller quadtree cells are used to communicate less uncertainty* (this was the model used in Kardos et al. 2005), *or is the opposite the case*? The question implies two metaphors of use for the *trustree*: (1) the metaphor of *detail* whereby the *trustree* subdivides more around accurate polygonal areas (attribute-wise), or (2) a new metaphor of *clutter* whereby the *trustree* tessellation is so dense that an uncertain polygonal area becomes covered or hidden from view (e.g. Figure 6.2).

The aim of this paper is to address the above question and related questions, such as: why would it be beneficial in the first place to determine if the *trustree* uncertainty method expresses a particular kind of metaphor? How does that metaphor communicate spatial and attribute uncertainty?

Does it communicate one type of uncertainty more effectively than the other? Does use of the metaphor produce a usable display?

Having been originally explored in a visual context by Bertin (1983) relating to his graphical variables, metaphors have been a prominent research area in cartography and geovisualisation. They figure highly in MacEachren and Kraak's (2001) geovisualisation agenda, as refined by Slocum et al. (2001) for cognition and usability issues. Juxtaposed with Fairbairn et al.'s (2001) representation challenge, our research is put into context with two main research goals:

1. *Metaphors* are an accepted method to develop a visual representation of spatial data, making a geovisualisation more effective (Slocum et al. 2001). Metaphors use some phenomena with which we are already familiar (*detail, clutter*) to explain something that is less understandable or tangible (*uncertainty*) (Petersen 1995). New metaphors based on *trustree* tessellation are presented in this paper to explore different conceptual theories about spatial and thematic uncertainty expressions. These are the opposing concepts of detail and clutter. In other words, does a viewer prefer to emphasise accuracy or hide inaccuracy?
2. A secondary question is that, having stripped away consideration of the two metaphors, is the *trustree* representation useful? There are resolution levels for *trustree* tessellations that will be more effective or usable than others. Ascertaining what these levels are is a representation issue.

For both research goals, the map stimuli are based on the New Zealand 2001 census of the Dunedin area, the population of which is represented using a choropleth map (with the census Area Unit being the areal unit mapped). The start of Section 5 has more details about choropleth classification and limitations. The uncertainty data from that census is represented using varying levels of transparent *trustree* tessellation, from low to high resolution. These variants were assessed for ability to communicate uncertainty on a metaphorical level, also their associated usefulness and usability (via an Internet survey).

The paper first defines attribute and spatial uncertainty in the context of this research programme. Next, a brief literature review selectively outlines current uncertainty visualisation methods and their metaphorical basis for visualisation. An account of the Internet survey and derived results is then given. Finally a discussion and conclusion to the paper are presented.

2 Uncertainty

Typically, uncertainty in its broadest sense exists in three forms – error, vagueness and randomness (Zhang and Goodchild 2002). It is error, which is the difference between a value obtained by measurement and the real value, and its opposite, accuracy, which is the type of uncertainty being addressed here. Nevertheless, uncertainty will be used as the name for error for the purposes of this paper, and has three components: spatial, attribute and temporal. Spatial information represented in a database is never truly accurate, because the data are a simplification of reality. Therefore, spatial uncertainty can arise from locational inaccuracy or the abstractions needed to force data into the prevailing raster and vector spatial data models. Temporal uncertainty occurs when the data inevitably becomes outdated. Finally, attribute uncertainty will always occur where and when an attribute is measured. These definitions have been formalised over the years (Hunter and Beard 1992, Goodchild 1994, Hunter and Goodchild 1995, Hunter and Goodchild 1997, Zhang and Goodchild 2002). Uncertainty is the major phenomenon that decision makers have to address when assessing data quality.

This research is concerned with two types of uncertainty – attribute and spatial. Temporal uncertainty does exist, but it is not a subject of this paper (see Tøssebro and Nygård 2003 for further detail).

2.1 Attribute Uncertainty

Attribute uncertainty (in a quantitative sense only) is the difference between collected data and the *actual* values exhibited in reality (note that no difference would mean total attribute accuracy). Goodchild (1995) explained that all attributes can be uncertain and may arise from numerous causes; from measuring instrument inaccuracies to human error. *Attribute uncertainty for census data* can come from different sources. As an example, data may be miscalculated; people may have their information collected twice or not at all. Some of these inaccuracies are unknown, some can be modelled while others can be directly measured and determined. Some census collection agencies attempt to gauge the amount of attribute uncertainty associated with a particular census. For example, Statistics New Zealand (2002) provides an additional post enumeration survey about the quality of the published results from the New Zealand 2001 census dataset and the more recent 2006 census (Statistics New Zealand 2006). The information contains modelled undercount figures and the assessment of other quantitative variables. While uncertainty is not known in spatial analysis for every application, its presence in the New Zealand 2001 census (being representative of a typical census dataset) underlies the uncertainty maps tested and reported on in this paper.

2.2 Spatial Uncertainty and Choropleth Maps

Spatial uncertainty is the locational difference between a representation of feature location stored in a geographic information system (GIS) and the *actual* location exhibited in reality (note also that no difference would mean total spatial accuracy). As some form of measurement, analysis and possibly generalisation are essential in translating a real world object's position into its digital equivalent stored in a GIS, some level of spatial uncertainty is inevitable. The research reported on in this paper deals indirectly with spatial uncertainty in that it is not concerned with the actual coordinate position of a spatial entity, more the spatial uncertainty associated with the subjectivity of census tract boundaries.

Census datasets are typically viewed using choropleth maps (Martin 1996) which lead to a number of spatial uncertainty issues, particularly when dealing with population data (Martin 1989; Chrisman 1989; Blakemore 1983; Jenks and Caspall 1971; Wright 1942). Uncertainty is seen to arise when aggregated data is assigned to areal units that are designed for ease of enumeration and administration, without regarding the characteristics of the data itself (for example, extreme socio-economic facets are likely to be misrepresented under these conditions – Martin 1989). Indeed, there are a potentially infinite amount of equally acceptable areal sizes and/or shapes under these assumptions, leading to an infinite number of differing patterns and values for a location (the modifiable areal unit problem – MAUP; Openshaw and Taylor 1981).

Homogeneity is also implied within geographic units (i.e. census tracts) and between units in the same category (i.e. as depicted by colour hue) (Jenks and Caspall 1971, Mark and Csillag 1989, MacEachren 1995), which would not be the case for semi-continuous population data. Pycnophylactic contours (Tobler 1979) and zone centroids (Martin 1989) are two possible techniques capable of managing these uncertainties.

Finally, uncertainty associated with choropleth class definition (actually an attribute uncertainty) is generated from the choices of techniques available (e.g. Natural Breaks - Alexander and Zahorchak 1943; Manual Definition- Jenks and Caspall 1971), each leading to different results.

In summary, choropleth spatial uncertainty encompasses: areal unit selection (size and shape), generalisation within areal units and choropleth class aggregation. Out of these, this research program considers expressions of uncertainty associated with areal unit size and shape selection and areal unit generalisation.

3 The Visualisation of Uncertainty

The importance of accuracy, data error and uncertainty have been known for some time (Goodchild 1992) and research into the visualisation of these intangible phenomena has pretty much always accompanied it. Several methods of visualising spatial and attribute uncertainty have been developed over close to 20 years, some of which will be reviewed here.

3.1 Visualising Attribute Uncertainty

MacEachren (1992) provided a good practical background to using Bertin's (1983) graphical variables in depicting attribute uncertainty. Specifically, MacEachren (1992) expressed that saturation of color (see also MacEachren et al. 1998; Leitner and Buttenfield 2000; De Cola 2002; Drecki 2002; Hengl et al. 2002; Aerts et al. 2003; Hengl et al. 2004), the focus of an image (e.g. obscuring fog – MacEachren 1992; McGranaghan 1993) and lowering spatial resolution might be effective ways to represent attribute uncertainty. The latter was proposed to convey uncertainty whereby the map graphic detail would change to correspond with the local attribute accuracy. Other methods put forward include map adjacency (MacEachren 1992; Schweizer and Goodchild 1992; Evans 1997; Leitner and Buttenfield 2000; Aerts et al. 2003), texture overlay (Monmonier 1990, Beard et al. 1991, MacEachren 1992), sound (Krygier 1994, Fisher 1994a, Fisher 1994b) and animation (Davis and Keller 1997, Ehlschlaeger et al. 1997, Bastin et al. 2002).

Out of the techniques listed above, adjusting resolution to visualise uncertainty is explored within this paper in a modified quadtree context (section 3.3). Furthermore, resolution is reinterpreted as a metaphor of detail. On the other hand, the clutter metaphor theoretically works in much the same way as obscuring fog in that it blocks the most uncertain areas from view. Another, perhaps closer analogue to clutter is the "dazzling" technique proposed by van der Wel et al. (1994). Discordant combinations of textures (line, pattern, orientation) and line would create an unpleasant as well as an obscuring effect to communicate uncertainty.

3.2 Visualising Spatial Boundary Uncertainty

Expressing spatial uncertainty for this research program generally means expressing uncertainty around fixed boundaries between spatial units. Techniques to show boundary uncertainty typically exploit the crispness or fuzziness of boundary edges such as the epsilon band (Perkal 1956) or confidence band (McGranaghan 1993). Though not specific to boundary uncertainty expression, Fisher (1994b) suggested the use of animation to show areas where spatial data are uncertain using individual map pixels that constantly change.

3.3 The Trustree to Visualise Attribute and Spatial Uncertainty

The *trustree* method tested in this research (Kardos et al. 2005) uses modified quadtree cell sizes to express attribute uncertainty. Choropleth boundary uncertainty is communicated qualitatively at the same time through the effective de-emphasis of that boundary by the overlaying quadtree cells (Figure 6.1). This method will be briefly explained as essential background to the research presented in this paper.

The specific data structure used, the Hexagonal or Rhombus (HoR) quadtree, was devised to recursively subdivide a bounded image into four equal-sized quadrates (Bell et al. 1989). Our initial research program proposed to take the HoR quadtree out of its original context and utilise its inherent variable resolution and graphical structure to exhibit the degree of attribute and choropleth spatial uncertainty (the more subdivisions the more certain the data beneath). Instead of using the HoR quadtree to decompose a bounded image array, it was used here to decompose census tract polygons in the following way:

1. A Monte Carlo simulation is run to generate uncertainty values for the census data (systematic error) and the modelled undercount for the study area (the information needed to parameterise the simulation is provided Statistics New Zealand; 2002). The two components are added and the error is propagated.
2. The user specifies the maximum number of divisions the quadtree should undertake, x.
3. The uncertainty data is then scaled between 0 (accurate) and x (uncertain) for easier construction of the uncertainty quadtree layer.
4. Decomposition takes place on the study area using the quadtree structure only if there is an accurate feature (i.e. a feature that has an uncertainty value less than the accuracy threshold, initially defined at (x-1),

for that level in the quadtree) at least 50% inside the quadrate. Otherwise the quadrate stops subdividing.
5. If subdivision takes place, the accuracy threshold decreases by 1;
6. Recursive division takes place until the final accuracy threshold level (0) is met. The finest resolution of the trustree is dependent on original map extent and number of divisions.

The visual use of the HoR quadtree in this way was termed the *trustree*. It was decided that the *trustree* should be represented as a transparent multi-resolution layer over the top of the choropleth map. This was inferred from results of a survey of transparency and other visualisation delivery methods reported on in Kardos et al. (2007).

As noted in section 3.1, the *trustree* theoretically synthesises the existing texture, resolution and fog methods of expressing attribute uncertainty, which is communicated through variable tessellation sizes. Communication of spatial boundary uncertainty lies in the utilisation of quadtree cells to distribute the boundary over space, reducing subjectivity and emphasis on the fixed boundaries associated with choropleth maps. However, there was an overriding question when using choropleth maps and *trustree* for uncertainty visualisation that this paper will attempt to answer: *how do humans mentally categorise the trustree output and what would an appropriate visual metaphor for the visualisation be?*

4 Metaphors

Metaphors are commonly used as an intuitive way to represent phenomena as, in complex cases, an effective metaphor can have an ability to explain reality better than reality itself. This it does "based on cross-domain correlations in our experience", leading to "perceived similarities" from which the metaphor gets its power (Lakoff and Johnson 2003, p.245), capable of operating on a cognitive, emotive and experiential level as well as the expected semantic and geometric communication (Lakoff 1987; Fabrikant and Skupin 2005). Indeed, it has been proved that metaphors are central and unavoidable in our lives, operating on a basic neuronal level (Lakoff and Johnson 2003). An example of a simple spatial metaphor is "time-is-proximity" as in the phrase "Christmas *is close to* New Year", where time is the target domain and proximity is the source domain (Lakoff and Johnson 2003).

4.1 Visual Metaphors

Metaphors can also work visually (as explored by Bertin, 1983) and it is on this basis that the *trustree* operates. Visual communication via metaphor depends on image schema working at a preconceptual level. Such schema include container, part-whole, up-down, centre-periphery, front-back, linear order and link schema (Lakoff 1987; MacEachren 1995). Importantly, these schema work in a spatial sense and will be explored in more detail in relation to the *trustree*.

The *trustree* uses or relates to many of the basic image schema listed in the previous section. The centre-periphery schema can be related to the invalid assumptions of the meshblock, that one value represents an entire area (ecological fallacy). For that meshblock, the value is likely to be true at the centre of the meshblock, with this likelihood decreasing towards the meshblock boundary, or periphery (this is the basis of Martin's [1989] use of meshblock centroids). The *trustree* conveys this to a limited degree, in that cells are more stable or of homogeneous size towards the centre of an underlying meshblock.

A linear order schema is implied in the spatial progression of cell size across boundaries, just as it is in play in the effective communication of the quantitative attribute scales (which in turn operate via an up-down schema) for population and uncertainty. Indeed, for the latter, more abstract concept (i.e. it has no physical form) the linear order schema is ideal as its means of communication (MacEachren 1995). Attribute uncertainty is effectively expressed to the viewer through these two schemas, which both serve to undermine the notion of a single value representing an entire meshblock.

The container schema works in two ways regarding the source census data used, as a container for census choropleth classes (attributes) and meshblock boundaries acting as a container for the meshblock area. By effectively diffusing such boundaries, the *trustree* seeks to address the limitations of the container schema in the second sense (that the basis of the boundary is arbitrary and that there should accordingly be no crisply-defined "inside" or "outside" of the container). In this way (also via the link schema explained next) spatial uncertainty is communicated in a qualitative sense to the viewer.

A link schema is implicit in the progression of cells across boundaries, effectively bridging meshblocks together. As MacEachren (1995) states (p.188): "…any locations on a map with a continuous symbol stretching out between them are assumed to be linked in some way". The fact that tessellations are by definition totally contiguous with no holes (Boots 1999) means that links are made across the choropleth irrespective of boundaries, therefore reducing their emphasis. The action of these major schemas in a *trustree* context is illustrated in Figure 6.3.

Other supporting schemas in operation with *trustree* display include those to do with the hierarchy present in the quadtree data structure (part-whole schema, up-down schema) and the front-back schema operating with the layered display of *trustree* and underlying choropleth map. That the *trustree* works via a variety of schemas reveals why it can be used to represent both spatial and attribute uncertainty.

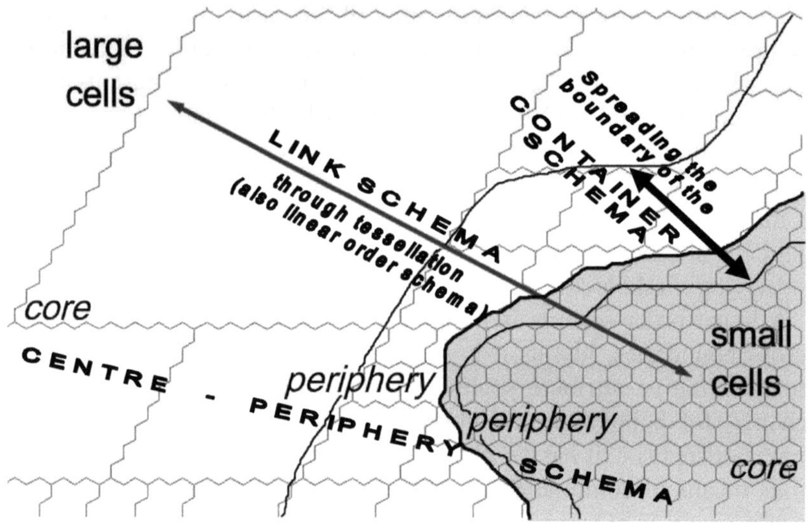

Fig. 6.3. Action of the three main image schemas relating to the *trustree* that operate with the two choropleth areas shown : container, centre-periphery and link schemas

In summary then, the container and link schemas address spatial boundary uncertainty but the centre-periphery and linear order schemas – while to a certain extent communicating spatial uncertainty – addresses the uncertainty of attributes. Even though only the single *trustree* tessellation is the basis, the schemas that operate through it seem to act separately for representing spatial and attribute uncertainty, being more powerful with the latter form. This is reinforced by the two metaphors that are associated with the *trustree* – detail and clutter – that have their metaphorical basis while communicating attribute, not spatial uncertainty. This is in agreement with Bertin (1983) and Lakoff (1987), who maintain that metaphors are most effective when used to convey a single concept. Nevertheless, this discussion suggests that there is a secondary metaphorical process that supports the communication of spatial uncertainty.

Some final thoughts on metaphors in general - the map in its own right is used as a metaphor to represent spatial and non-spatial data, using the basic image schema already discussed in relation to the *trustree*. Fairbairn et al. (2001) discussed using the map metaphor in terms of abstraction and noted that where data are realistic, a metaphor with a low degree of abstraction is chosen and vice versa. This would lead uncertain data (an abstract data concept) towards the abstract end of the representation spectrum. Dent (1993) suggested using regular metaphors for abstract representations and proposed using geometric shapes like circles, squares and triangles. Accordingly, the technique tested here is based on another simple geometric shape, the hexagon, as a metaphorical representation for attribute and spatial uncertainty (also chosen for visual appeal and representational accuracy – Kardos et al. 2003).

4.2 A Metaphor for the Trustree

Some common visual metaphors for visualisation of uncertainty techniques are expressed in Table 6.1. By defining such metaphors, a foundation for further analyses into new metaphors can be established. In particular, this paper is focused specifically on using metaphors evoked by hierarchical tessellations and choropleth displays.

It is proposed that the *trustree* can exhibit two commonly-understood metaphors – 1) *levels of detail*, or 2) *levels of clutter* or working on a basic level as "uncertainty is sketchy" (or "uncertainty is coarse-grained") and "uncertainty is a barrier" respectively. Both metaphors are exhibited in Figure 6.2, whereby the form of the circle is seen through the well-spaced lines, but when the lines become too dense the circle becomes cluttered over.

The *trustree* structure's variable cell size can show many levels of detail (with less detail equal to more uncertainty). There is an implication when viewing many cell sizes on the same map that more is known about the finer resolution cells when juxtaposed with a coarser resolution cell. The comparatively empty space in the latter implies that less is known or less detail is available about the area covered. This situation is enabled by the quadtree data structure, operating through a part-whole schema. The detail metaphor is akin to the resolution metaphor (MacEachren 1992; Leitner and Buttenfield 2000), but rather than changing the original map (as in de Bruin 2004) the *trustree* becomes a transparent layer lying on top of the unchanged map.

Table 6.1. Visualisation of uncertainty techniques and their associated metaphors

Technique	Reference	Certain Areas	Uncertain Areas	Metaphor
Display By Value (Map Adjacency)	e.g. Evans 1997	Low Values	High Values	Uncertainty is dark
Saturation of Color	e.g. McEachren et al. 1998	High Saturation	Low Saturation	Uncertainty is impure
Pixel Mixture	De Gruijter et al. 1997	Single Hue	Multiple Hues	Uncertainty is confusion
Dazzling	Van der Wel et al. 1994	Clear / Low amount of differing patterns	High amount of differing patterns	Uncertainty is unpleasant
Texture Overlay	e.g. Monmonier 1990	Small Texture Amounts	Large Texture Amounts	Uncertainty is a barrier
Fog Overlay	e.g. McEachren 1992	Clear	Foggy	Uncertainty is a barrier
Image Sharpness	e.g. McGranaghan 1993	Sharp Focus	Blurry	Uncertainty is hard-to-see
Resolution	e.g. McEachren 1992	Fine Resolution	Coarse Resolution	Uncertainty is coarse-grained
Blinking Areas	Fisher 1994b	No Blinking	Blinking On Areas	Uncertainty is unstable
Sound	e.g. Krygier 1994	Low Pitch	High Pitch	Uncertainty is loud

The clutter metaphor works on the same physical setup but reduces the transparency to communicate uncertainty (the obscuring fog metaphor works by the same barrier metaphor and has latterly been termed by transparency instead of fog – MacEachren 1995). Practically it works by subdividing the quadtree more with increasing uncertainty and the metaphor would come into effect when the collective representations of at least some of the cell boundaries in the *trustree* visualisation are sufficiently dense to block underlying information from view. Put succinctly, this research tries to answer the question of whether the *trustree* is more effective on a cognitive level at revealing certain areas of hiding uncertain areas.

5 The Experiment

An Internet survey was created to assess a number of variations on the *trustree* visualisation, designed to determine which metaphor is more appropriate when utilising the *trustree* to express uncertainty; detail or clutter. For the underlying census data (see section 2.1 for details of this dataset), the population data was classified by the Quantile method into six classes. The colour schemes used for the choropleth maps were consistent between visualisations, chosen by ColorBrewer (Brewer et al. 2003).

Although it is not advisable to depict absolute values in a choropleth map (e.g. Kraak and Ormeling 1996) it is employed here to permit a simple expression of published data and its uncertainty component (omitting error propagation tasks that would arise from standardisation by area). However, it can be argued that a gross error is needlessly introduced by using the choropleth in this way. While true, it is not a major issue in the context of this paper, where emphasis is on uncertainty visualisation and its metaphorical expression (though the challenge of legitimising an absolute value choropleth through an associated and essential visual uncertainty expression – where the uncertainty arises from using absolute values – is an intriguing one).

To generate values for attribute uncertainty on this dataset (i.e. the data conveyed by the *trustree*), the census undercount data was modelled using Monte Carlo statistical methods (see Kardos et al. [2005] for more details).

5.1 Survey Visualisations

For the survey, the choropleth census map was used, with a corresponding population legend, overlaid by a transparent *trustree* tessellation representing uncertainty (with no legend containing explanatory or distracting numbers, to place the emphasis on the tessellation). The choropleth census map had a consistent background whilst the overlaying *trustree* tessellation varied. The first visualisation (Figure 6.4a) used a large cell pattern where the census tracts were easily observable through the tessellation. As the visualisations progressed, the resolution of the smallest cell increased until the sixth visualisation (Figure 6.4f), when the smallest tessellation clearly cluttered over census tracts. The first survey hypothesis is –

- The initial visualisations (Figure 6.4a & b) exhibit a metaphor of detail and the final visualisations (Figure 6.4e & f) exhibit a metaphor of clutter.

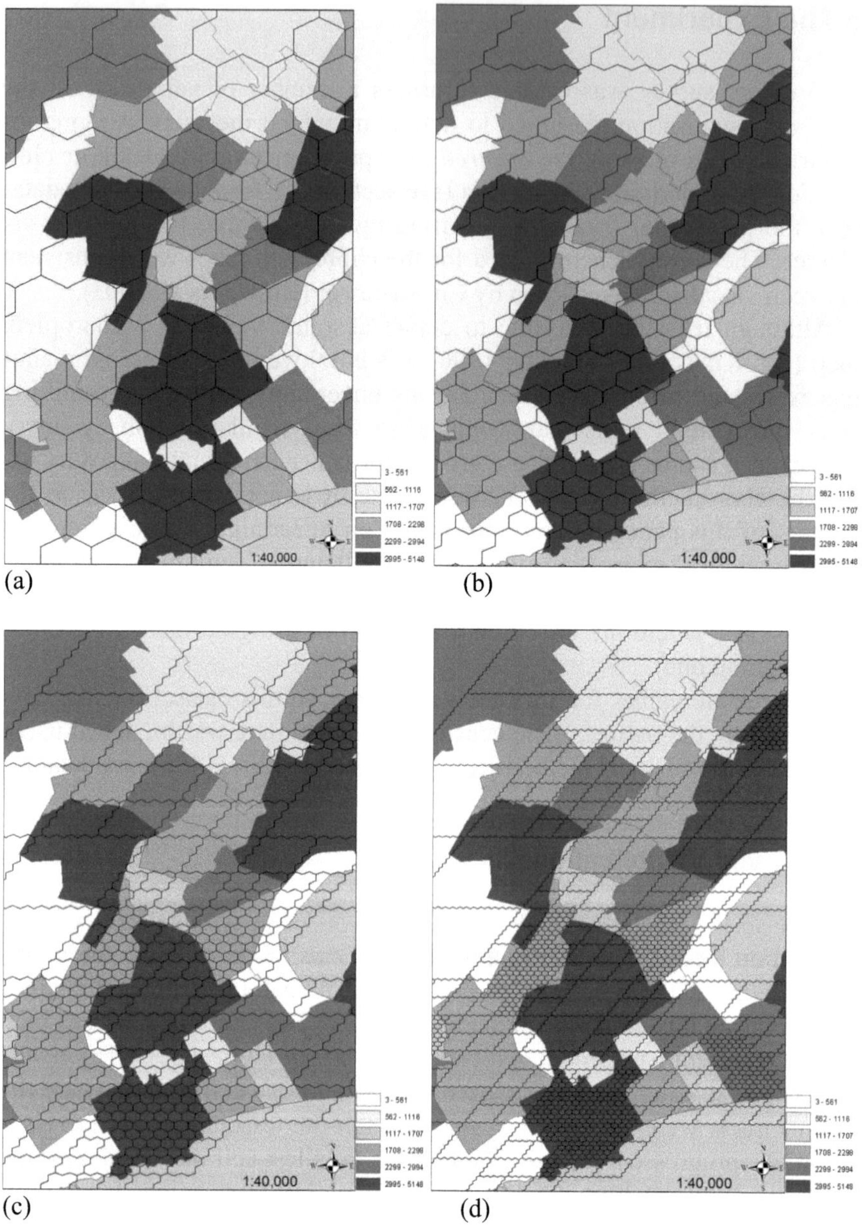

Fig. 6.4. Census population choropleth visualisations of Dunedin city census tracts with varying resolutions of the *trustree* tessellation overlay. (a) – Visualisation using the *trustree* tessellation at the lowest resolution, through to (f) – visualisation using the *trustree* tessellation at the maximum resolution (e and f on next page).

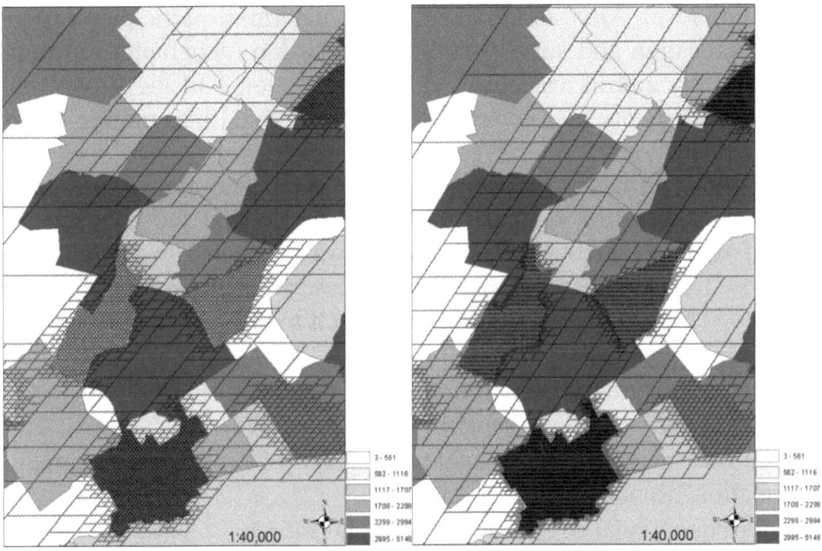

(e) (f)

Fig. 6.4. (cont.)

By collecting usability results simultaneously, the survey was designed not only to determine the appropriate metaphor to use, but also the most usable tessellation level and the metaphor exhibited at that level. The established metaphor (determined by its implementation in previous surveys) is that of detail and it is being challenged by the clutter metaphor as a possibly more dominant facilitating factor in communicating uncertainty. Therefore a secondary hypothesis is –

- The clutter metaphor is the most appropriate metaphor for communicating uncertainty, as defined by viewer opinions on how usable it is.

The *trustree* tessellations and underlying choropleth map were generated in ESRI ArcGIS using Visual Basic script (therefore implementable in a commercial environment). The dual layer maps were then exported as a bitmap for use in the Internet survey.

5.2 Survey Questions

Two similar surveys were devised to complement each other. The questions and format were consistent, but the resolution level for some images was slightly different. Choropleth spatial boundary uncertainty and attribute uncertainty, as explained in the context of this paper, was provided for participants to read. The results from both surveys were combined based on resolution level. After an initial section to determine participant experience, specialisation and survey-specific knowledge (geographic information, uncertainty and its visualisation; this is Section 1 of the survey) the choropleth uncertainty maps were shown, one-by-one (this, and the rest of the survey is Section 2).

Along with the maps, a definition of the detail and clutter metaphors was provided, allowing each participant to understand the concepts involved. If required, additional information was available to explain how the *trustree* tessellation could express a measure of attribute and spatial uncertainty. Three questions accompanied each *trustree* visualisation (Figure 6.5):

- Does the overlaying tessellation depict a metaphor of detail or clutter? (binary answer – *detail* or *clutter*)
- Do you find this tessellation usable as a means of expressing uncertainty information? (binary answer – *yes* or *no*)
- Free format optional comment on the visualisation.

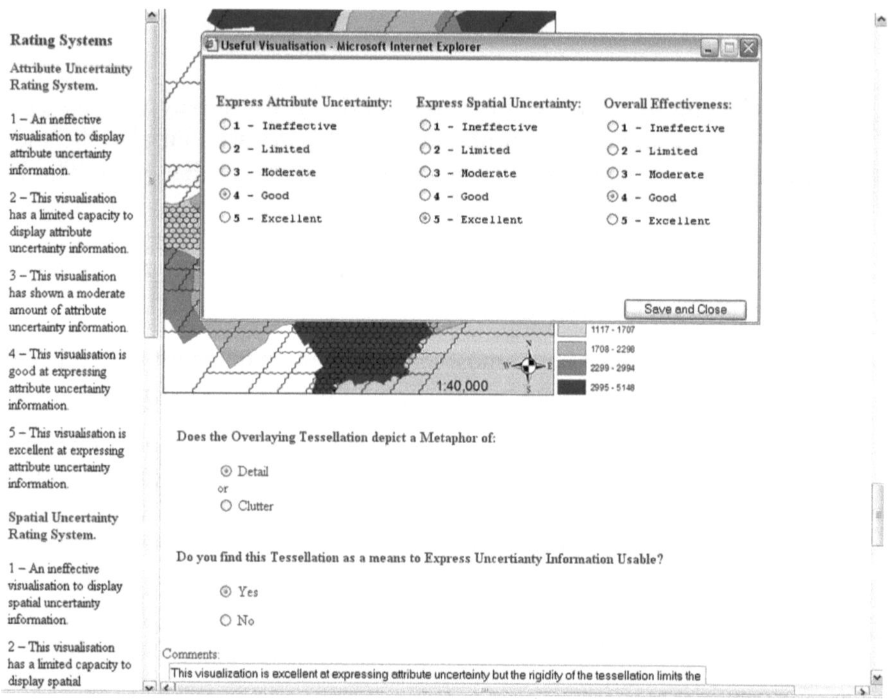

Fig. 6.5. A screenshot of the Internet survey. The attribute, spatial and overall rating schemes are found on the left. Two questions and a commentary box follow the visualisation. Depending if the visualisation is usable, a different popup box would appear. In this image the tessellation was selected as usable to express uncertainty, therefore the participant was requested to rate it.

If the participant selected the *Yes* option for the second question, he/she was prompted to rate the visualisation for expressing *Attribute Uncertainty*, *Spatial Uncertainty*, and *Overall Effectiveness* (Figure 6.5). The user was asked to rate the visualisation on a 1-5 scale whereby:

1 – Ineffective, 2 – Limited, 3 – Moderate, 4 – Good, 5 – Excellent

An in-depth explanation of the rating system was provided to the participants. If the user selected the *No* option, deciding that the visualisation was not usable in expressing uncertainty information, then a separate prompt would open. The prompt gathered information about why the visualisation was *not usable*. Five check boxes were presented: *Don't Understand*, *Confusing*, *Complicated*, *Uncertainty Information Unreadable*, and *Other*. If *other* was selected then a free format text box was provided. At the end of the six visualisations the user was asked:

- Which metaphor would you choose if the tessellation was used to visualise uncertainty information?

Invitations to participate in the survey were sent to GIS mailing lists. 82 complete responses were received. It was decided to run the survey over the Internet as despite the acknowledged pitfalls (Klassen and Jacobs 2001; Fowler 2002) it had clear logistical advantages over controlled experiments (accessibility to skilled participants), mail surveys (more cost-effective) and telephone surveys.

5.3 Survey Results and Analysis

From the first section of the survey, there was a broad, evenly distributed range of participant experience with using geographic information. All but a few participants used geographic information for viewing / presenting and the overwhelming majority were aware of uncertainty in GI. The favoured methods of uncertainty representation were by using colour and texture as graphical variables.

The Section 2 (of the survey) results were converted into percentages, as shown in Figure 6.6. The visualisations, based on their resolution level, will be referred to as *a*, *b*, *c*, *d*, *e* and *f* from now on. Figure 6.6a shows the preference results for overlaying *trustree* tessellation using a metaphor of detail or clutter, for all six resolution levels. Error bars have been provided at a 95% confidence level. The perception of a detail metaphor is initially low in visualisations *a* & *b* and then gradually increases to a peak in *d* (i.e. the detail metaphor dominates) and then returns to a low level when the tessellation level becomes too dense in *f*. The inverse of these results are seen for clutter, where *a* & *b* are clearly cluttering the display, and again as the tessellation becomes too dense around *e* and *f*.

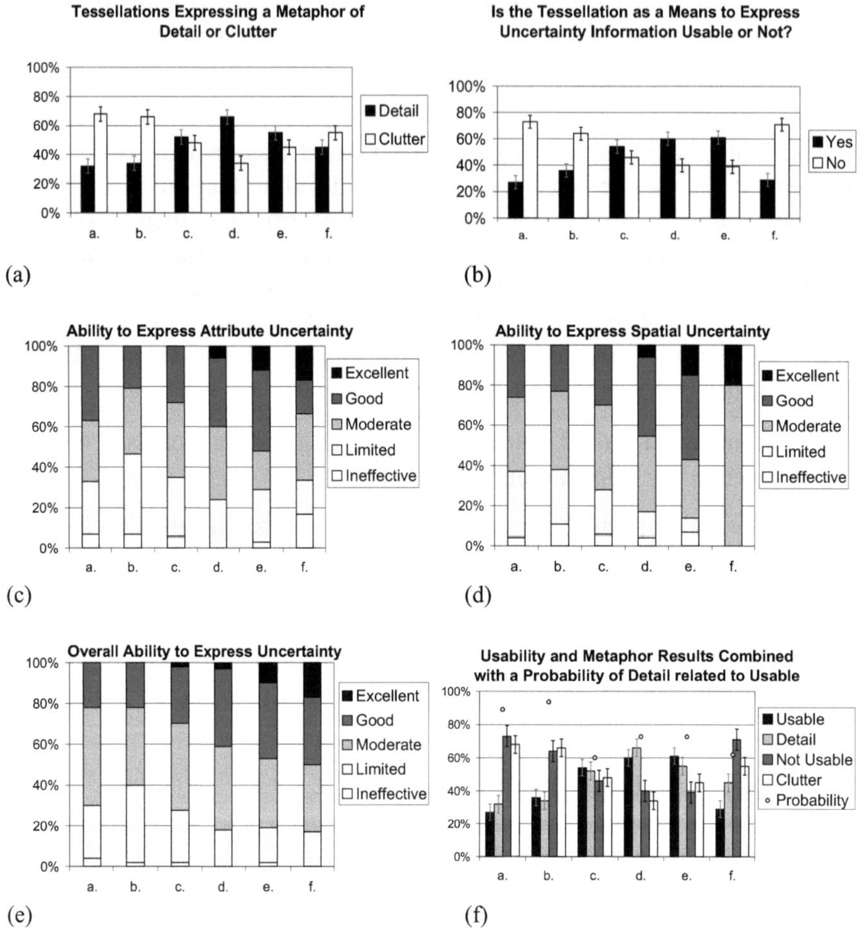

Fig. 6.6. Survey results for Section 2 – Determining an Appropriate Metaphor for six resolutions of the *trustree* tessellation, which correspond to Figure 7.4a – 7.4f.

Figure 6.6b shows results from *a* - *f*, stating if the *trustree* tessellation was able to express uncertainty information in a usable manner. The initial visualisations of *a* & *b* are not able to express uncertainty information by majority, but as denser tessellation becomes apparent in *c, d* & *e*, the visualisations become usable (60% for *d*; 61% for *e*). When the tessellation level became too dense to differentiate cells in *f*, the usability was markedly reduced to 29%.

If a participant selected *yes* (corresponding to usable) for a particular tessellation, a rating was requested for attribute and spatial uncertainty.

There are similar trends between the attribute uncertainty, spatial uncertainty and overall ability graphs (Figure 6.6c, 6.6d & 6.6e respectively). These graphs all show a gradual progression in the ability to express uncertainty information, as more resolution is added through the tessellation. With a 95% confidence interval, all results have 8.9% margin of error. Chi square tests established that, with the exception of the results for attribute uncertainty (significant at the 95% level only) and overall uncertainty (not significant), all results were significant at the 99% level.

The last survey question asked participants if they would use the *trus-tree* tessellation as a metaphor of detail or clutter. Detail dominated, with 89% of the participants making it their visual metaphor of choice. This was confirmed in Figure 6.6f where the usability and metaphor results become merged into one graph. There was an obvious relationship where detail and usability form a trend (falling at a maximum of 6% from each other) for all visualisations (except *f* which has a 16% separation). The same relationship was seen between clutter and 'not usable'. The probability results of detail matching usable were plotted. Only in *c, d* & *e* is the tessellation usable, and detail is the dominant metaphor to use for these visualisations. Chi square tests statistically confirmed the affinity of detail and usability and the non-affinity of clutter and usability.

Following each of the visualisations *a* – *f*, was the free format question. For *a*, many comments were that the tessellation cell size was so large that it did not provide enough information to be usable. Also, it was commented that there was no context or legend given to determine the uncertainty information. This information was deliberately omitted to reduce bias. If a legend explained that a smaller area was more accurate then it would immediately bias the tessellation towards a metaphor of detail and vice versa. Comments for *b* explained that the visualisation was more appropriate than *a*, but the tessellation could interfere with the underlying data. In *c* & *d* participants commented that the tessellation could provide greater detail and how it interfered less with the display but was still understandable. The fine scale geometric pattern tended to contrast enough with the underlying data allowing the two datasets to be seen without interference. The small tessellation size was interpreted as having more accurate information in that area, but the different sized polygons were hard to compare. Also, the finer tessellation could demonstrate uncertainty at the *edges* or spatial uncertainty better than the coarse tessellation. In *e* and *f* contributors stated that the tessellation had the *look* of conveying more information (i.e. detail) and intuitively representing varying levels of uncertainty. The obscuring effect of clutter in *f* was also commented on.

The hypothesis at the start of this section stated that most of the participants would choose a metaphor of detail for *a* and *b*, and a metaphor of clutter for *e* and *f*, with ramped results between. However, participants considered *a* and *b* to clutter the images with a metaphor of detail being evident for *c*, *d* and *e* (only *f* was classified as clutter, like the hypothesis). Therefore the hypothesis should be refuted, overall. The secondary hypothesis which states that visualisations exhibiting a metaphor of clutter will also be the most usable displays is also refuted on the presented evidence.

6 Discussion and Conclusions

A visualisation of uncertainty technique, the *trustree*, has been tested for its ability to communicate uncertainty on a metaphorical level. The *trustree* purports to reveal the spatial boundary uncertainty and attribute uncertainty inherent in choropleth maps. At a preconceptual level, the *trustree* (which is based on the quadtree) uses areal tessellations to undermine the crisp boundaries of the container schema implicit in the choropleth. The same continuous tessellations use a link schema to bridge choropleth boundaries, further highlighting their invalidity. At the same time, attribute uncertainty is communicated through the cells by introducing a centre-periphery schema to the hitherto homogeneous choropleth polygon. Finally, a linear order schema conveys some sense of the gradual change in attribute level uncertainty that occurs across a choropleth boundary.

However, prior to this experiment it was not clear how the *trustree* worked metaphorically as a whole, whether finer cells were indicative of less uncertainty, implying more knowledge / detail about an area (this detail or "uncertainty is sketchy" metaphor works on a basic resolution structure) or whether the same cells worked to obfuscate the underlying data if sufficiently dense (the clutter or "uncertainty is a barrier" metaphor suggested elsewhere by use of transparency). That both detail and clutter gain their metaphorical power while communicating attribute uncertainty led to the insight that spatial uncertainty was being relayed visually through a less effective metaphorical means. This also agrees with previous research that accounts for the specific action of a (visual) metaphor to convey a single concept.

An Internet survey designed to address this problem ascertained the following:

1. When results for the detail metaphor was paired with those for usability, there is at least a 95% chance that they correlate. The inverse was seen for the clutter metaphor and "not usable". Therefore, it could be

said that when the *trustree* is shown in a usable fashion, a metaphor of detail is exhibited.

2. Contrary to the first presented hypothesis, the low resolution tessellations (i.e. *a* & *b*) rated highly as clutter, probably because the pattern was not dense enough to provide any graphical meaning and may have visually interfered with the choropleth boundaries. Therefore they were considered as meaningless clutter over the map display and not usable. On the other hand, as denser tessellation became apparent (*c*, *d* and *e*) the ability for patterns to be seen through the tessellation, providing detail and usefulness to the display, increased too. But, as the tessellation became too dense (*f*), the patterns were still apparent but cluttered the display to a point where detail was no longer seen and the visualisations became unusable. This aspect at least agreed with the first hypothesis. However, the secondary hypothesis, that visualisations that exhibit the clutter metaphor are the most usable has to be refuted.
3. A parallel is seen between differing resolution tessellations and typical cartographic maps. Scale and feature generalisation on a cartographic map is used to make a map readable, and not clutter the display. If too much detail is provided on a cartographic map the display becomes cluttered, just as the tessellation in *e* & *f* did. So, in a sense *clutter* is where a map feature does not match the scale at which the map is being displayed. Also, if the *trustree* tessellation was too sparse or too dense there was a lack of usability arising from a mismatch in generalisation. Moreover, clutter as it is expressed here, could be considered an inefficient generalisation process. As with cartographic maps, the *trustree* tessellation resolution should convey the information at an ideal level (or scale) for maximum ability to read and use.
4. There was very little difference between the effectiveness of metaphors to convey spatial and attribute uncertainty until visualisation *e* & *f*. Here, the results imply that a cluttering tessellation can convey spatial uncertainty better than attribute uncertainty (which contradicts the supposition that attribute uncertainty is the dominant form of uncertainty being communicated). This could be due to a smaller tessellation suggesting a smooth diffusion of boundaries, whilst the larger tessellation shape seems more discrete and does not diffuse well with irregular census tract boundaries.

Further research is required to assess the worth of the *trustree* in a day-to-day decision making environment that utilises spatial representations. The prototype implementation tested here is implemented in a widely-available GIS software package, which would facilitate adoption of the display. However, there are issues of training non-specialist decision makers (in interpreting maps of geographical uncertainty) to overcome. The user training and assessment could be part of a larger initiative covering a suite of existing visualisation of uncertainty methods. The approach outlined above is in tune with the current overriding concern that future research emphasis should be on the context or application rather than the visualisation technique itself (e.g. Hope and Hunter 2007a). Finally, further research is required to determine whether other spatial data models beyond the HoR quadtree are effective in expressing uncertainty or even if the *trustree* is applicable to other standard maps outside of the choropleth (e.g. the cartogram).

Acknowledgements

This research would not be possible without funding from the School of Business at the University of Otago, New Zealand. Concurrent research was undertaken at L'Institut d'Administration des Entreprises (IAE), Université Jean Moulin Lyon III, hosted by Laïd Bouzidi. Finally, we also appreciate the useful comments that the reviewers have provided to improve this paper.

References

Aerts JCJH, Clarke KC, Keuper AD (2003) Testing Popular Visualization Techniques for Representing Model Uncertainty. Cartography and Geographic Information Science 30(3):249-261

Alexander JW, Zahorchak GA (1943) Population-Density Maps of the United States: Techniques and Patterns. Geographical Review 33(3):457-466

Bastin L, Fisher P, Wood J (2002) Visualizing Uncertainty in Multi-Spectral Remotely Sensed Imagery. Computers and Geosciences 28(3):337-350

Beard MK, Buttenfield BP, Clapham SB (1991) NCGIA Research Initiative 7: Visualization of Spatial Data Quality. National Center for Geographic Information and Analysis, Technical Paper 91–26, Santa Barbara, CA, pp 1-175

Bell SBM, Diaz BM, Holroyd FC (1989) The HoR Quadtree: An Optimal Structure Based on a Non-Square 4-Shape. In: Brooks SR (ed) Mathematics in Remote Sensing. Clarendon Press, Chelmsford, Oxford

Bertin J (1983) Semiology of Graphics. The University of Wisconsin Press, Madison, Wisconsin

Blakemore M (1983) Generalisation and Error in Spatial Data Bases. Cartographica 21(2&3):131-139

Boots B (1999) Spatial tessellations. In: Longley P, Goodchild M, Maguire D, Rhind D (eds) Geographical Information Systems, vol 1. John Wiley, pp 503-526

Brewer CA, Hatchard GW, Harrower MA (2003) Colorbrewer in Print: A Catalog of Color Schemes for Maps. Cartography and Geographic Information Science 30(1):5-32, http://www.colorbrewer.org/

Chrisman NR (1989) Modelling Error in Overlaid Categorical Maps. In: Goodchild M, Gopal S (eds) The Accuracy of Spatial Databases. Taylor & Francis, London, pp 21-34

Davis TJ, Keller CP (1997) Modelling and Visualizing Multiple Spatial Uncertainties. Computers and Geosciences 23(4):397-408

De Bruin S, De Wit AJW, Van Oort PAJ, Gorte BGH (2004) Using Quadtree Segmentation to Support Error Modelling in Categorical Raster Data. International Journal of Geographical Information Science 18(2):151-168

De Cola L (2002) Spatial Forecasting of Disease Risk and Uncertainty. Cartography and Geographic Information Science 29(4):363-380

De Gruijter JJ, Walvoort DJJ, van Gaans PFM (1997) Continuous Soil Maps - a Fuzzy Set Approach to Bridge the Gap between Aggregation Levels of Process and Distribution Models. Geoderma 77(2-4):169-195

Dent BD (1993) Cartography : Thematic Map Design. Wm C Brown Publishers, Dubuque, Iowa

Drecki I (2002) Visualisation of Uncertainty in Geographic Data. In: Shi W, Fisher PF, Goodchild MF (eds) Spatial Data Quality. Taylor & Francis, New York, pp 140-159

Dykes J, MacEachren AM, Kraak M-J (eds) (2005a) Exploring Geovisualisation. Elsevier, London

Dykes J, MacEachren AM, Kraak M-J (2005b) Advancing Geovisualization. In: Dykes J, MacEachren AM, Kraak M-J (eds) Exploring Geovisualisation. Elsevier, London, pp 693-703

Ehlschlaeger CR, Shortridge AM, Goodchild MF (1997) Visualizing Spatial Data Uncertainty Using Animation. Computers and Geosciences 23(4):387-395

Evans BJ (1997) Dynamic Display of Spatial Data Reliability: Does It Benefit the Map User? Computers and Geosciences 23(4):409-422

Fabrikant SI and Skupin A (2005) Cognitively Plausible Information Visualization. In: Dykes J, MacEachren AM, Kraak M-J (eds) Exploring Geovisualisation. Elsevier, London, pp 667-690

Fairbairn D, Andrienko G, Andrienko N, Buziek G, Dykes J (2001) Representation and Its Relationship with Cartographic Visualization. Cartography and Geographic Information Science 28(1):13-28

Fisher P (1994a) Hearing the Reliability in Classified Remotely Sensed Images. Cartography and Geographical Information Systems 21(1):31-36

Fisher P (1994b) Animation and Sound for the Visualization of Uncertain Spatial Information. In: Hearnshaw HM, Unwin DJ (eds), Visualization in Geographical Information Systems. Wiley & Sons, Chichester, West Sussex, England, pp 181-186

Fowler FJJ (2002) Survey Research Methods, 3rd edn. Sage, London

Goodchild MF (1992) NCGIA Research Initiative 1 Accuracy of Spatial Databases. University of California, Santa Barbara, pp 1-9

Goodchild MF (1994) Spatial Accuracy. In: AURISA, Proceedings of the 22nd Annual Conference, Australasian Urban and Regional Information Systems Association, Sydney, pp 1-9

Goodchild MF (1995) Attribute Accuracy. In: Guptill SC, Morrison JL (eds) Elements of Spatial Data Quality. Elsevier, New York, pp 59-80

Hengl T, Walvoort DJJ Brown A (2002) Pixel and Colour Mixture: GIS Techniques for Visualisation of Fuzzyness and Uncertainty of Natural Resource Inventories. In: Hunter G (ed) Proceedings of the International Symposium on Spatial Accuracy in Natural Resources and Environmental Science, 10-12 July 2002. Melbourne, Australia, p 271

Hengl T, Walvoort DJJ, Brown A, Rossiter DG (2004) A Double Continuous Approach to Visualization and Analysis of Categorical Maps. International Journal of Geographical Information Science 18(2):183-202

Heuvelink GBM, Burrough PA (2002) Developments in Statistical Approaches to Spatial Uncertainty and Its Propagation. International Journal of Geographical Information Science 16(2):111-113

Hope S, Hunter GJ (2007a) Testing the effects of positional uncertainty on spatial decision-making. International Journal of Geographical Information Science 21(6):645-665

Hope S, Hunter GJ (2007b) Testing the effects of thematic uncertainty on spatial decision-making. Cartography and Geographical Information Science 34(3):199-214

Hunter GJ, Beard K (1992) Understanding Error in Spatial Databases. The Australian Surveyor 37(2):108-19

Hunter GJ, Goodchild MF (1995) Dealing with Error in Spatial Databases - a Simple Case Study. Photogrammetric Engineering and Remote Sensing, 61(5):529-537

Hunter GJ, Goodchild MF (1997) Communicating Uncertainty in Spatial Databases. Transactions in GIS 1(1):13-24

Jenks GF, Caspall FC (1971) Error on Choroplethic Maps: Definition, Measurement, Reduction. Annals of the Association of American Geographers 61(2):217-244

Kardos J, Benwell GL, Moore A (2007) Assessing Different Approaches to Visualise Spatial and Attribute Uncertainty in Socioeconomic Data using the Hexagonal or Rhombus (HoR) Trustree. Computers, Environment and Urban Systems 31(1):91-106

Kardos J, Benwell GL, Moore A (2005) The Visualisation of Uncertainty for Spatially Referenced Census Data Using Hierarchical Tessellations. Transactions in GIS 9(1):19-34

Kardos J, Moore A, Benwell GL, (2003) The Visualisation of Uncertainty in Spatially-Referenced Attribute Data Using Trustworthy Data Structures. In: Whigham P, Moore A (eds) SIRC 2003. Dunedin, New Zealand, pp 11-25

Klassen RD, Jacobs J (2001) Experimental Comparison of Web, Electronic and Mail Survey Technologies in Operations Management. Journal of Operations Management 19(6):713-728

Kraak M-J, MacEachren A. (2005) Geovisualization and GIScience. Cartography and Geographical Information Science 32(2):67-68

Kraak M-J, Ormeling F (1996) Cartography: Visualisation of Spatial Data. Longman, Dorchester

Krygier JB (1994) Sound and Geographic Visualization. In: MacEachren A, Taylor DRF (eds) Visualization in the Modern Cartography. Pergamon, New York, pp 149-166

Lakoff G. (1987) Woman, Fire and Dangerous Things: What Categories Reveal About the Mind. University of Chicago Press, Chicago

Lakoff G, Johnson, M (2003) Metaphors We Live By, 2nd edn. University of Chicago Press, Chicago

Leitner M, Buttenfield BP (2000) Guidelines for the Display of Attribute Uncertainty. Cartography and Geographic Information Science 27(1):3-14

MacEachren A, Brewer CA, Pickle LW (1998) Visualizing Georeferenced Data: Representing Reliability of Health Statistics. Environment and Planning A 30(9):1547-1561

MacEachren A, Kraak M.-J (2001) Research Challenges in Geovisualisation. Cartography and Geographic Information Science 28(1):3-12

MacEachren AM (1992) Visualizing Uncertain Information. Cartographic Perspective 13(Fall):10-19

MacEachren AM (1995) How Maps Work: Representation, Visualization, and Design. Guilford Press, New York

MacEachren AM, Robinson A, Hopper S, Gardner S, Murray R, Gahegan M, Hetzler E (2005) Visualizing Geospatial Information Uncertainty: What We Know and What We Need to Know. Cartography and Geographical Information Science 32(3):139-160

Mark DM, Csillag F (1989) The Nature of Boundaries on Area-Class Maps. Cartographica 21(1):65-78

Martin D (1989) Mapping Population Data from Zone Centroid Locations. Trans Institute British Geographers 14(1):90-97

Martin D (1996) Geographic Information Systems: Socioeconomic Applications, 2nd edn. Routledge, London

McGranaghan M (1993) A Cartographic View of Spatial Data Quality. Cartographica 30(2&3):8–19

Monmonier M (1990) Strategies for the Interactive Exploration of Geographic Correlation. Proceedings of the 4th International Symposium on Spatial Data Handling, IGU:512-521

Openshaw S, Taylor PJ, (1981) The Modifiable Areal Unit Problem. in: Wrigley N, Bennett RJ (eds) Quantitative Geography: A British View. Routledge and Kegan Paul, London

Perkal J (1956) On Epsilon Length. Bulletin de l'Academie Polonaise des Sciences 4(3):399-403

Petersen MP (1995) Interactive and Animated Cartography. Prentice-Hall, New Jersey
Schweizer DM, Goodchild MF (1992) Data Quality and Choropleth Maps: An Experiment with the Use of Color. Proceedings GIS/LIS 2, pp 686–699
Slocum TA, Blok C, Jiang B, Koussoulakou A, Montello DR, Fuhrmann S and Hedley NR (2001) Cognitive and Usability Issues. In: Geovisualisation. Cartography and Geographic Information Science 28(1):61-75
Statistics New Zealand (2002) A Report on the Post Enumeration Survey 2001. Statistics New Zealand, Wellington
Statistics New Zealand (2006) A Report on the 2006 Post-enumeration Survey, http://www.stats.govt.nz/analytical-reports/report-2006-post-enumeration-survey.htm (accessed 7th May 2008)
Tobler W (1979) Smooth Pycnophylactic Interpolation for Geographical Regions. Journal of the American Statistical Association 4(367):519-536
Tøssebro E, Nygård M (2003) A Discrete Model for Spatial Data with Temporal Uncertainty. In: 2nd IASTED International Conference on Information and Knowledge Sharing. Scottsdale, AZ, USA, pp 57-68
van der Wel FJM, Hootsmans RM, Ormeling F (1994) Visualization of Data Quality. In: MacEachren AM, Taylor DRF (eds) Visualization in Modern Cartography. Pergamon Press, Oxford, pp 313-331
Wright JK (1942) Map Makers Are Human: Comments on the Subjective in Maps. Geographical Review, 32(4):27-544
Zhang J, Goodchild MF (2002) Uncertainty in Geographical Information. Taylor & Francis, London, New York

CHAPTER 7

Simulation and Representation of the Positional Errors of Boundary and Interior Regions in Maps

Tomaž Podobnikar[1,2]

[1]Scientific Research Centre of the Slovenian Academy for Sciences and Arts, Ljubljana, Slovenia

[2]Institute of Photogrammetry and Remote Sensing, Vienna University of Technology, Vienna, Austria

Abstract

The main goal was to describe the positional errors in maps. Although the errors of spatial data are mainly normally distributed, their character often contains admixtures of poorly explained heterogeneities and other uncertainties that are not simply described. Better understanding the error distributions requires performing quantitative and qualitative tests. Portions of the error distributions that can be explained with stochastic behavior were modelled by Monte Carlo simulations. Two principles for the evaluation of projected error models and types were applied: boundary and surface simulation. Both principles address random, locally systematic, and systematic (with the limits at the gross) error distributions. The boundary error was simulated with vector lines and polygons. The surface error was simulated by producing error surfaces that shift every grid point or square-shaped polygon. Visualisations of the simulations include fuzzification process and allow additional understanding of the nature of errors. The study sets were historical maps and land use data derived from them. Additionally we propose a strategy for reconstructing data models and the error models dependent on them.

Keywords: historical maps, quality, positional error, fuzzy boundaries, Monte Carlo simulation, land use

1 Introduction

The quality of spatial data depends on many elements. Data are traditionally evaluated using well-defined and standardised data models. The data models are relatively comprehensible for topographic maps because of the long traditions of surveying and map production. Preliminary (a priori) quality assessment of the data, especially of maps, requires a high degree of skill, addressing factors such as the original purpose of the map, its lineage, usage, etc. This knowledge can include information about mapping scale, methodology, generalisation methods, possible media (paper) deformation, the precision and conscientiousness of surveyors, and the mathematical basis. Actual (a posteriori or subsequent) data evaluations verify or disprove the preliminary assessment of quality. Reference data sets, which suppose no error, present a data model which is compared with actual data. Both data sets should be in the same reference coordinate system. The problem with evaluating positional quality, especially of maps, is a lack of temporal quality, therefore the reference points for evaluation should be chosen very carefully.

The data researched in this study are historical maps, where the data model and the character of quality parameters are more complex, and include subjective factors and uncertainties. The study should account for space and time; consider the historians, geographers, surveyors, and cartographers; and support the reconstruction of a physical and human landscape (Plewe 2002). Acquired information can stimulate the imagination and increase the understanding of possible quality parameters of data and error distribution models (Peterca et al. 1974; Burrough and McDonnell 1998). At the same time, we should be aware that it is rarely possible to obtain unique and objective enough metadata information, which is even more important for understanding historical maps than for contemporary data. Potential uncertainties include:

- The quality of historical trigonometric networks; the possibilities of mapping with “eyeballing”

- The role of relevance, importance (military secrets, strategic areas), and obstacles (swamps, mountains, state borders, war fronts, bad weather conditions, etc.) in the pattern of mapping
- Different and unknown generalisation methods, in spite of common rules
- The motivation, skill and character of operators (measurer, data processor) and the effect of costs

Statistical techniques (numerical or quantitative) have traditionally been employed for quality control and error assessment, according to common standards. Visual (graphical or qualitative) methods are less accepted; nevertheless they can be important tools for both preliminary and subsequent evaluation (Wood and Fisher 1993; Burrough and McDonnell 1998). These methods are excellent for obtaining first impressions and other insights about data sets such as historical maps. Unfortunately, they require an operator who has an expert knowledge of the studied landscape to improve objectivity. Visual methods may play an important role in evaluation of old maps, especially in combination with quantitative and other qualitative methods.

We propose an alternative strategy for quality control that includes visual and simulation methods. This is because little information about the data sources (old maps) is available. The methodology begins with classical digitisation and georeferencing, then applies standard statistical quality control and analysis techniques. An important part of the process is the acquisition of as much information as possible about data sources and their quality, using mostly qualitative methods. This leads to a possible hypothesis of one or more error distribution models that are then simulated with Monte Carlo (MC) statistical methods. This combination of simulation and visualisation may leads to significantly better insight into the data source, enabling us to establish a clearer error distribution due to the specified (or chosen) data model. It can lead to improvements in georeferencing and more reliable spatial analyses in GIS where complex homogenisation procedures for fusion of data from different quality historical maps to the unique data model are required.

As the proposed procedures of this research have an iterative nature, we consider the optimal approach to be an appropriate combination of deductive (hypothesis first, data later) and inductive (hypothesis later, data first) approaches. This combination is referred to as an abductive approach. This allows for interaction between data exploration and human (operator) perception, and the discovery of patterns along with the hypothesis (Anselin 2005). A full understanding of the quality of background and error distributions is fundamental for obtaining reliable simulation results. This methodology stresses MC error simulation and its visualisations using fuzzy classification. Although it can be applied to any data sources, some methods of this research were adapted especially for the Triglav national park case study region in Slovenia (Fig. 7.1). There we tried to determine the quality and usability of a range of historical maps, used for land use dynamics analysis (Podobnikar 2009).

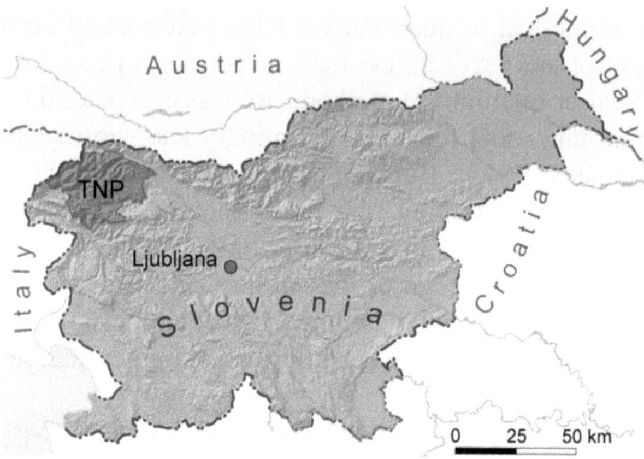

Fig. 7.1. Location of Triglav national park

2 Theoretical Background

The connection between the real world (geographical reality) and the data set can be schematically described as: real world – data model (model quality) – data (data quality). The data model or nominal ground is a conceptualisation and representation or abstraction of a real world that is a selected representation of space, time, or attributes (Aalders 1996). The model quality is a semantic quality of representation by which a complex reality, not quantitatively measurable, is acquired. Data are described by the type of spatial objects to which variables refer and by the level of measurement of these variables. Data quality refers to the performance of the data given the specifications of the data model, where deficiency in data quality results from errors in the data. The overall relationship between the real world and the data can be described as the uncertainty (of the mapping), addressing both model quality and data quality (Haining 2003).

The assessment of data quality from the user or producer perspective is expressed in terms of how closely data values represent a chosen data model for a corresponding reality (Blakemore 1984). The producer should be able to describe the data quality and make available a comprehensive quality report for the product, while the user or decision maker should know how to formulate and specify requirements for the data. The producer is usually an expert in data acquisition and spatial data processing, but is typically not skilled with applications to spatial data. In contrast, the data user is often an expert in a particular area such as cartography, history, or environmental modelling, but has a limited knowledge of data handling or quality assessment. For communication between them, it is reasonable to imply a standardised language; the interaction may be supported with visualisations of the problem (Clarke and Teague 1998). Unfortunately, the great potential of visualisation as a qualitative approach to the quality assessment is too often neglected in the praxis and the uncertainty of the data is not well transmitted from the academic to the users' sphere. The standardised data model can increase interoperability between users and producers, minimise cost and increase the quality. Additionally, the methodology for description of the elements of data quality should be precisely defined (Martinoni and Bernhard 1998).

International standards distinguish (ISO/TC 211, project 19113) five elements of data quality: positional, temporal, and thematic accuracies, completeness, and logical consistency. Accuracy can be defined as a positional, temporal, or thematic difference between a value from the data and the corresponding value as it appears in the data model. Furthermore, we can distinguish absolute from relative accuracy. The position of objects could be assigned to absolute accuracy and the irregularity of the shapes of objects to relative accuracy (that is, relative to general position). The term accuracy is considered to mean absolute accuracy if not explicitly specified. Another aspect of the term accuracy is referred to as precision. Relative accuracy can be treated as precision when measurements are independently repeated. It is related to the scale, resolution, or generalisation of the measurements. The precision of data increases when the particular values approach one another more and more closely as one makes more and more measurements more and more carefully. The term precision describes a type of error that can not be considered as a mistake or something that can be corrected. Accuracy can be depicted as a value taking the form 11.15 ± 0.06, where the true value according to the data model is expected to lie within the interval of 11.09 to 11.21, with a probability of 67% assuming normal distribution. In contrast, precision can be conveyed by the number of decimal places in the data (Khorram et al. 1999), for example 11.15 vs. 11.1524. In other words, accuracy is described as a deviation of the derived value from the actual one and precision is the dispersion of multiple measured values of the same phenomenon (Buckner 1997). In popular terms we can say that accuracy is an attempt to state the truth, but precision is the ability to tell the same story (not necessary the truth) over and over again. All data contain uncertainty as a consequence of inaccuracies inherent in the process of taking measurements (Haining 2003). Uncertainty can never be completely eliminated from mapping data.

2.1 Positional Error

Error is considered to be either a lack of data quality or data with little or no accuracy. Positional accuracy describes one aspect of quality. Other elements of quality, poorly understood with respect to their importance, nature, etc., are difficult to consider as part of positional quality because the limits between them are not clear. There are other minor views (see Taylor and Kuyatt 1994) that formalise the qualitative and quantitative concepts of the term positional error. Different types of positional error attribute accuracy to precision and even to systematic or gross error. Additionally we should consider the possibility that a random component of error may become systematic and vice versa under different error models or after Brown and Heuvelink (2007) uncertainty models. Unfortunately a systematic component that is handled in this study can not be easily detected and removed with, for example, mean error. The errors are therefore statistically, less and even nonstatistically describable. The last are not standardised concepts that are therefore difficult to analyse. Different errors or effects can be recognized and assessed by employing effective methods of quality control.

The most extreme error is defined as gross, denoting a mistake in measurement or an experiment that has been corrupted in some way. Indicators for gross errors can be statistical outliers (in some views also residuals) but these values may not be wrong. Outliers that are extreme with respect to the overall distribution of (attribute) values can be distributional in nature. The gross error for these outliers may be suspected from a histogram plot, but there may also be spatial outliers, where the values may be extreme given their spatial position relative to the neighbouring values.

We can distinguish two groups of errors: measurement (inherent) and processing error (spatial analyses). Possible sources of inherent errors include: the source or lineage of the data, the completeness of data sources, mapping scale, temporal discordance, digitisation, sampling density and significance (digitisation is really just sampling), georeferencing, transformation between data formats, accessibility and price, etc. (Bernhardsen

1992). The errors can be evaluated by statistical methods or by other means (Taylor and Kuyatt 1994). Regarding these definitions, accuracy can be described with a root mean square error (RMSE), and precision as a standard deviation or standard error (σ). Processing errors cause error propagation and depend on the types of analyses and on the associated data structures. Measurement errors may be propagated to map or other spatial data sets. Analytical error distributions can be more complex than numerical ones. A mismatch in spatial (or temporal) correlation structures may be one of the greatest problems when combining data from different sources or from different phenomena or attributes. The most well-known error propagation method is a first-order Taylor series approximation (Taylor and Kuyatt 1994; Heuvelink 1998).

Numerical simulation methods such as the MC are appropriate for more complex models of error evaluation and description, where we don't know much about the character of inherent or processing errors and why they occur (Burrough and McDonnell 1998). They assume spatial continuity and spatial homogeneity of error. The simulations are relatively easy to carry out and are generally applicable. A disadvantage is that the results do not come in a nice analytical form. The equation resulting from the Taylor methods may be easily analysed, to determine how reduction of input error will affect the output, but with the MC method, the only solution is to run the entire simulation process again. Another disadvantage of the MC method is the numerical load, which requires extensive computational time for complex computational models. The main advantages of the MC simulations are that they are very flexible and can yield the entire distribution.

2.2 Simulation of the Positional Error

MC simulations as statistical methods are generally used in cases where physical processes are random or the theoretical mathematics in the different hypothesis tests is weak (Openshaw et al. 1991; Brown and Duh 2004). The MC methods of positional error analysis follow the statistical theories of error distribution and propagation (Burrough and McDonnell 1998). For example, an error model may be applied to a digital elevation model (DEM) and then a simulated viewshed (see Fig. 7.2; Fisher 1994, 1996), as well as slope, aspect or drainage errors. Similarly may be simulated path courses between chosen fixed points based on a least-cost algorithm (Ehlschlaeger et al. 1997). Errors can be simulated to evaluate differing qualities of classification of borders between (land use) classes. The simulations may also follow spatial analyses that provide combinations of land use, slope, aspect, sun exposure, tourism resources, etc. A complex probability model can be specified for every source of error if the understanding of spatial variability components is high. Multiple and equally probable outcomes from the model are often needed because a single simulation of the random function/fields is really just one of a large number of representations of the specified probability model (Cressie 1993; Haining 2003).

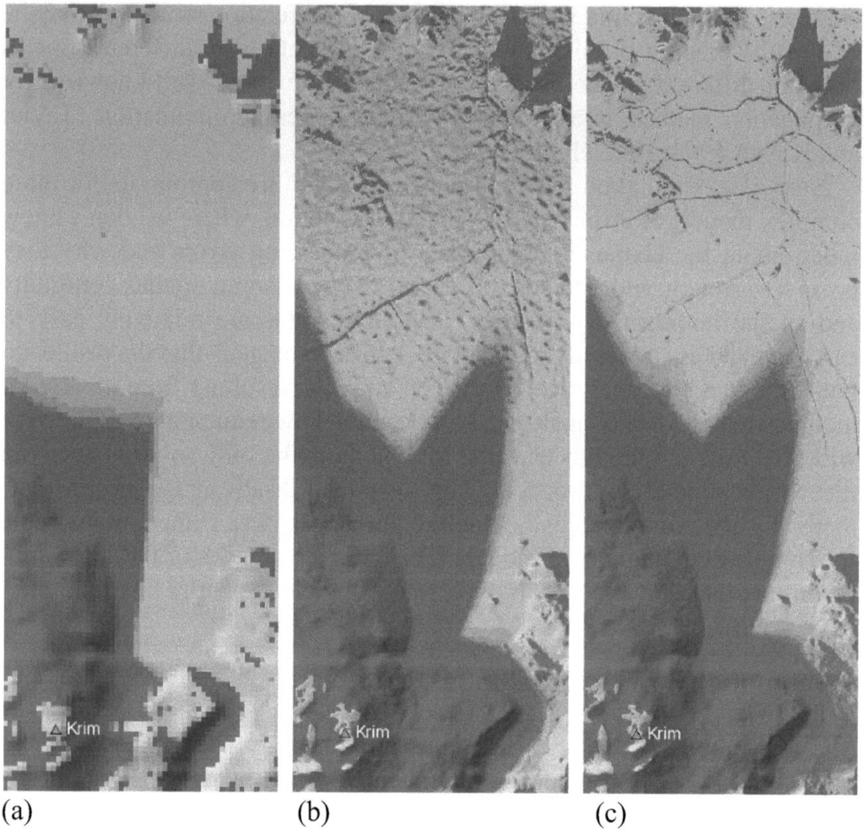

Fig. 7.2. MC simulations that produce a probability viewshed from the observation point (Krim), applying error models (continuously varying error distribution surfaces) to the evaluated quality of DEMs with a resolution of (a) 100 m; (b) 25 m – interferometric radar (IfSAR); and (c) 20 m. The probability viewshed was converted to fuzzy viewshed by semantic import model (Burrough and McDonnell 1998). The borders of viewsheds are therefore fuzzy. The inherent error model for the oldest DEM (a) is underestimated. Red indicates shadows, with a lower possibility of visibility (as regards the five fuzzy classes). At the top are hill shadows of tested DEMs (area of 6 by 16 km). Similar effects may be obtained with positional error simulation of the observation point.

One outcome (realisation) of the random function can be generated using unconditional simulation. That yields outcomes which are consistent with the probability model but in which simulated data values do not correspond with known data values. In the case of a normal (Gaussian) distribution, the mean value (μ) is 0 and the standard deviation (σ) is 1. Simulations that respect data are conditional. In this case they come from a normal probability model where mean and distribution are specified but at the locations where data values are known the outcomes match these. Transformation of the unconditional uniform distribution of a discrete random variable via cumulative to normal distribution can be applied by the Box-Muller method (Box and Muller 1958). This is a suitable approach for simulation of error in spatial data. Variables used for MC simulations of error are usually spatially autocorrelated with respect to the nature of the error. A general procedure for the MC simulation algorithm is then:

1. Generate a set of random numbers ($i = 1, 2, 3, \ldots$, n)
2. Transform n random numbers (using the Box-Muller method) to an appropriate unconditional distribution (normal probability model)
3. Respect current data and the associated error model to compute a conditional distribution; store individual output
4. Repeat steps 1 through 3 N times
5. Analyse and evaluate a distribution from the N outputs

The number of simulations depends on the objectives of the study and on the surveyed phenomena. In some cases even one outcome can be enough to assess the spatial variability of a stationary random function. In the case of non-stationary random functions, hundreds and even more outcomes should be processed (Kanevski and Maignan 2004). N = 100 outputs for the positional error simulation of land use boundaries is considered as sufficient irrespective of further fuzzification of the boundaries with appropriate raster resolution chosen and considering standard deviation values σ.

2.3 Fuzzy Description and Presentation of Error

Positional errors caused by incorrect coordinates can be represented with simplified objectivity by so-called ε- or epsilon-bands (Perkal 1956; Blakemore 1984; Giordano and Veregin 1994; Fisher 1996; Molenaar 1998) which describe error zones around the boundaries, where the outer line is considered to be σ. As we proposed, a more realistic presentation can be provided by MC simulation methods interpreted by fuzzy sets. The examples (a-d) in Figure 7.3 present a probability distribution model for a line segment (Shi et al. 1999) that is converted to a fuzzy set (e).

The fuzziness of geographical objects can also be applied to features such as the boundaries of polygons, or to variation in membership function values, interpolated from point data. A continuously varying surface as a fuzzy field can be inferred. These more complex problems are described in the following sections.

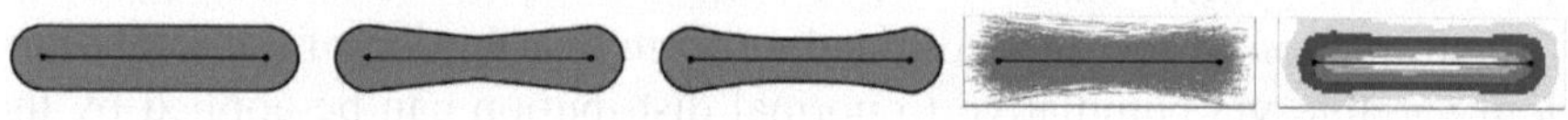

Fig. 7.3. Two points bound the line. Positional error is presented with different approximations of the epsilon-bands or –zones (a–e), and simulated with MC method (d–e) where (e) presents visualizations using fuzzy classification

3 Materials and Methods

For this case study, the Josephine Ist Military Topography (JMT28.8, 1:28 800) was used dating back to the end of 18th century (Rajšp and Ficko 1995). These Austrian map sheets are considered the first mapping of the entire territory of Slovenia. Additionally, the following data sets were statistically analysed to better evaluate the process of determining the quality of the mapping: Italian (ITK25), Yugoslav (TK25) and Slovenian maps of scale 1:25 000, and Yugoslav (TK50) in scale 1:50 000, dating from the beginning to the end of the 20th century. Analysis of a smaller area is based on orthorectified mid-20th century aerial photographs (AP,1:15 000) and Franciscan cadastral maps (FC2.8, 1:2880 to 1:5760) from the beginning of the 19th century. A contemporary land use data set was acquired from orthophotos and a land use data set (Fig. 7.4).

There were some reasonable drawbacks to the historical maps, which were considered in building the error distribution model. The oldest maps were found in the archives and only paper prints (unfortunately not copper plates) were available for scanning to a digital format. Therefore substantial deformations of the paper could be expected that were not easily distinguished from deformations of triangulations and other sources. Different maps were made with different mapping techniques, varying object catalogues (map legends) and a lack of projection and transformation parameters. For example, the JMT28.8 was available only on paper sheets, which were photographed and scanned for our use. The origins of the cartographic projections were based on Paris, Rome, Greenwich, the Ferro Island, and local origins. In general we might choose between two basic possibilities for transformation to a common coordinate system: known parameters or reference points. For the oldest maps, an original reference system was not available, so we used reference points.

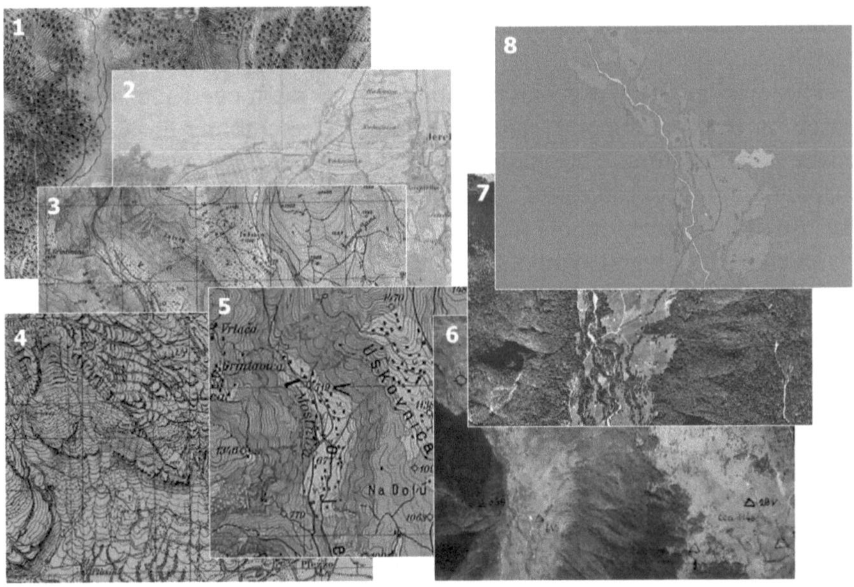

Fig. 7.4. Historical data sets: 1-JMT28.8, 2-FC2.8, 3-TK50, 4-ITK25, 5-TK25, 6-AP, 7-orthophoto, 8-land use; the example is from Voje and Uskovnica in the eastern part of the Triglav national park (area of 4 by 3.2 km)

The next problem is illustrated by typically sparsely surveyed mountain areas resulting in less accurately made maps, especially among the oldest of them. Many details were not measured or were mapped just by eyeballing, and therefore succumbed to gross errors. Nevertheless, identical points on the historical maps, linked with contemporary orthophotographs (or better contemporary maps of similar scale), were carefully chosen considering historical knowledge of measurement techniques and possible environmental changes. Geometrically, the best points were the trigonometrical ones. These were mostly churches or towers. Much confidence was put into the positions of possibly triangulated bridges, crossroads, castles, summits, etc., as well to some river confluences and distinctive meanders that were possibly well positioned and stable throughout the last centuries. The older maps are, the less confidence can be placed in identical points, due to natural or anthropogenic changes in land use. Identical points were sometimes difficult to identify due to toponyms in Slovene, German, and Italian languages. For the oldest maps, it was better to match identical points roughly – even for poorly detected details, in order to more systematically cover an entire mapped area and avoid locally large distortions. Even if the oldest maps look poor in their positional quality, they are very valuable for their toponyms, buildings, roads, etc., all important to understanding a cultural landscape and the overall quality of the (old) data model.

The original land use classifications were rearranged into 14 classes based on the contemporary land use data set, resembling the Corine land cover nomenclature. The majority of classes were distinguishable on all maps, though in some cases, just barely. Difficulties were also caused by the use of different symbols for the same categories, for example, using aerial texture or polygon symbols for the same phenomena. The problem was the most apparent in the mosaicking of the TK25 on the Yugoslav and the ITK25 on the Italian side, and was solved using the less detailed TK50 from the same period which covers the entire study area. The ITK25 used textures, so the borders between different land uses are very blurry and therefore more difficult to digitise to sharp polygons. The overall accuracy of the digitised land use from the ITK25 was ultimately the worst – even worse than that of the JMT28.8! For better categorisation, additional maps, data sets and information were used: the geomorphology of relief elevation, aspect, proximity of urban areas, the increasing elevation of the upper vegetation level during the last decades, and the general order of vegetation zones. Additionally, the borders between the classes were less precisely mapped and they are much more difficult to interpret than the carefully selected identical points used for georeferencing. Land use data for every historical period was acquired by a backwards editing method – map-by-map, starting with the newest vector-based land use data.

The difficulties described above greatly affected the thematic and positional accuracy of the georeferenced maps and derived land use data. They were evaluated on the digitised land use data sets despite the difficulty in some cases of distinguishing them. Another problem was the extrapolation of the general positional error of entire map surfaces to the boundary error in classified polygons, representing polygons of land use. This could be also considered as a thematic or attribute problem. We tested the thematic accuracy of classified land use with the standard PCC and Kappa tests. Briefly, we can say that most of the error comes from the less precisely mapped land use features in the original maps and positionally/thematically incorrect interpretations of the boundaries between the classes. Other mostly empirical tests were applied and are described in Podobnikar (2009). From this point it makes sense to leave possible difficulties that can occur during digitisation of land use from the original historical maps for other research discussions. We considered that after a sensitive georeferencing of the maps (not digitised land use data), the error distribution model was related only to positional accuracy, evaluated with (see also Table 7.1):

- RMSE, evaluated using different methods of transformation through georeferencing
- AME (averaged maximum error), calculated as differences between manually selected identical points on georeferenced maps and nominal data sets (contemporary orthophotos and maps), and then averaged; as areas of the maps were chosen with obviously lower expected quality, the AME was considered to be locally maximal; this measure controls RMSE numerically as well as visually by marking distances and directions

Table 7.1. Positional accuracy evaluation of historical data sets (as appeared in the maps, in mm)

	JMT28.8	ITK25	TK25	TK50	FC2.8	AP
RMSE[a]	16.7–19.1	1.7	0.2	1.5	3.8–4.5	0.2
AME[b]	19.8	1.7–2.5	2.0–2.4	1.8–2.0	4.9	0.4–0.5
Max.	36.5	4.8	5.4	7.5	6.9	0.8

[a]Tested for 250 to 1600 reference points per map sheet
[b]Tested for 10 to 60 reference points per map sheet

By statistical and visual evaluation of historical maps and an exploration of the methodology and principles of historical mapping, we deepened our knowledge and understanding of the overall quality of the maps and the positional error distributions of acquired land use data sets. Many of the objective and subjective tests that were applied are described in a greater detail in Podobnikar (2005, 2009). For example, one of the easiest methods of error evaluation is the transparent overlaying of discovered data sets with the highest quality (nominal) data set. Distorted areas can be easily assessed visually and possible error distributions indicated.

3.1 Positional Error Classification for Simulation

The sheets of the oldest map, JMT28.8, were the subject of a detailed analysis. The basic conditions for establishing a complex model, based on objective and subjective factors, were fulfilled for simulations and later visualisations. Error distribution models include many components of uncertainty that can be handled either implicitly or explicitly. With greater understanding of data quality, we can more explicitly define particular variables and the error distribution model for MC simulations. Due to possible errors in the most complex map, JMT28.8, the simulations of the land use data were performed with the following assumptions:

- random (mostly relative) effects: instruments, human factors and others that are indefinable with respect to precision
- locally systematic (mostly relative) effects over larger areas that could be also classified as spatially smooth or random on a smaller scale: triangulation errors, errors in measurements of long distances, avoiding obstacles, absence of measurements, other indefinable errors, distributed as described
- systematic/gross (mostly absolute) effects over the whole case study area: projections, triangulation problems, measurements

Assumptions for the second and the third categories of error were not as easily and uniquely simulated as those for the first category were. Systematic or even gross errors may not be distributed randomly, but their nature is sometimes similar to stochastic, and at the same time they are almost impossible to remove. In spite of the inexact and uncertain classification of error types, we attempted to simulate all three groups using statistical MC simulation as the most practical but still reasonable solution. In any case, many more error distributions can be found with further detailed examination. In all likelihood, most of them could not be evaluated by statistical methods, and in the worst cases not even with quantitative approaches.

Technically, the MC error simulation of error models and types was applied using two different principles that were denoted as the "first" and "second" principle. The proposed methods of simulation involved several common steps that were considered in both principles. The first step was to create random values applying standardised unconditional normal distributions. Those distributions were multiplied with the RMSE values of the JMT28.8 (Table 7.1). The RMSE was transformed to standard deviation values σ that are suitable for simulations of stochastic processes with the following simplification: RMSE = σ. The RMSE came from real measurements between the data models (contemporary orthophotos and maps) and the data sets (JMT28.8), but with σ, possible deviations from the JMT28.8 were simulated, using the RMSE for reference. With a minor generalisation, σ as precision could be a good approximation for RMSE as accuracy. We considered that the actual degree (value) of RMSE is possibly higher than σ, but the character of error was modelled consistent with previous tests. All simulations were handled with ESRI's macro language AML, but processed with GNU C.

3.2 Vector Boundary Simulation Method (First Principle)

The first approach under the first principle followed the rules of boundary simulations (Burrough and McDonnell 1998) applying vector lines. RMSEs measured on the JMT28.8 were assigned to the land use boundaries acquired from the JMT28.8 as σ. The average σ was 28 800 · 17.9 mm = 515 m (Table 7.1)! Even though this is high, we were not able to reduce this error, as more reference points could not be located. The additional uncertainty of σ was simulated with σ_c, where $\sigma_c \ll \sigma$ and the final $\sigma' = \sigma \pm \sigma_c$. Numerically, the positional accuracy varied from 480 to 550 m in a real world scale of 16.7 to 19.1 mm in the data set (map) according to different larger homogeneous areas. Therefore $\sigma' = 515 \pm 35$ m.

Furthermore, we assumed that the shape of a particular simulated object remains similar to the acquired original. Thus the objects should still be recognizable after the simulation. Applying that condition, neighbourhood nodes of the boundary lines should be correlated. Rather than programming the correlation between the nodes within the simulation of one data set, a simpler approach was to apply the correlation coefficient over the entire study area. That solution also handled some of the above-mentioned assumptions of systematic/gross errors. The correlation coefficient CORR was applied using Eq. 7.1. The chosen value of CORR depends on correlation between generalisation of the vector lines and σ, or, on the other hand, on the computed ratio of absolute to relative accuracy. The numerical evaluation was ascertained visually by the example presented in Figure 7.5 where the appropriate CORR is between 0.5 and 0.75. With regard to the unconditional normal distributions, x_1, y_1, were calculated once for entire data set, and x_2, y_2 separately for every node point. The pair x_1, y_1 estimates a mostly systematic component, while x_2, y_2 a more random one.

$$\begin{aligned} x_{\text{changed}} &= x_{\text{orig.}} + \sigma'(x_1 \cdot \text{CORR} + x_2(1-\text{CORR})) \\ y_{\text{changed}} &= y_{\text{orig.}} + \sigma'(y_1 \cdot \text{CORR} + y_2(1-\text{CORR})), \text{CORR} \in [0,1] \end{aligned} \tag{7.1}$$

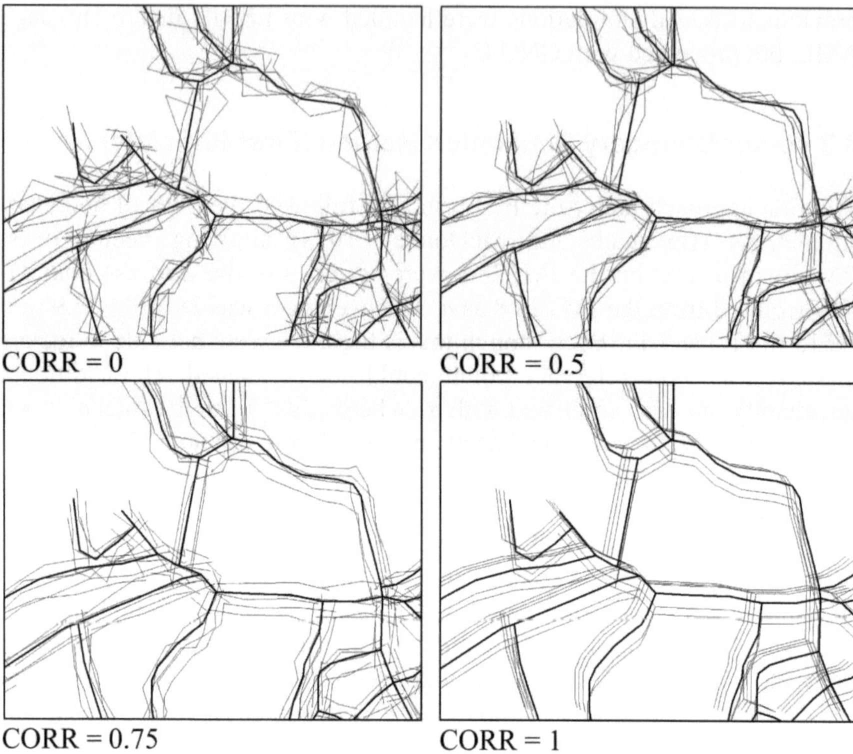

Fig. 7.5. MC with N = 5 simulations of the vector lines boundaries using different correlation coefficients between neighbour nodes. The bold line represents the original data

The second approach under the first principle was analogous to the first, except that this simulation was applied to the boundaries between polygons and not to the lines themselves. To preserve the original topology as much as possible after changing positions of the nodes on simulated polygons requires a much more complex numerical solution. Our second approach handles this problem quite adeptly, even where low correlation coefficients had been used. Random effects were simulated for both line and polygon approaches. With CORR between 0.5 and 0.95, we also simulated some systematic effects. The simulations were repeated for N = 100 times and the singular data sets were rasterised to unique resolution, and then all summed up. The last step was to convert the summed up the probabilistic data by fuzzy data sets and visualisation in several different ways:

- lines: by presenting N = 5 simulated data around original lines (see Fig. 7.5)

- lines: fuzzy boundary as a “shadow” effect using all (N = 100) simulated lines (see Fig. 7.6)
- polygons: fuzzy boundary by presenting N = 100 simulated data using different colours for dissimilar classes of fuzzy sets; false colour RGB composites is suggested based on the colour mixture technique (Hengl et al. 2002; see Fig. 7.7 where composition of red and blue is applied)

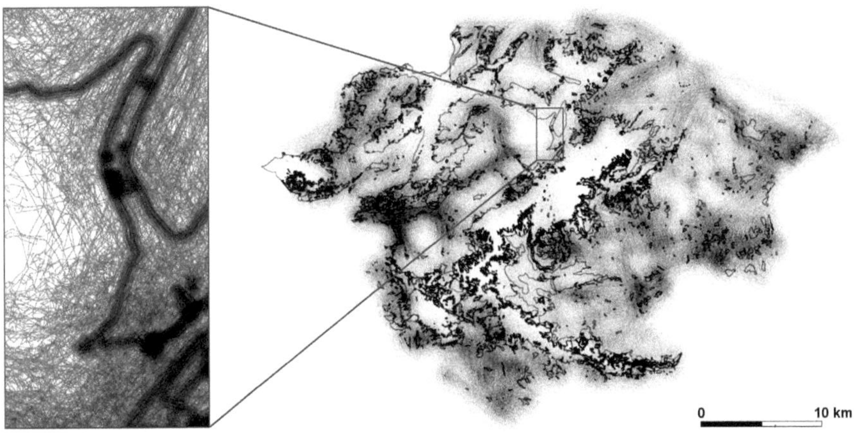

Fig. 7.6. Land use data acquired from the JMT28.8 simulated N = 100 times with the MC method using the land use boundaries for the entire Triglav national park (area approx. 50 by 50 km). Distributions of the borders show possible error expanses. Some are less diffused than others depending on the accuracy of the land use with σ applied. It can be clearly seen that land use is uncertain for more than 90% of the area

3.3 Continuous Surface Simulation Method (Second Principle)

With continuous surfaces, the error distributions were simulated for entire areas of land use acquired from the JMT28.8 maps. The idea arose from the determination that important parts of error distribution are locally systematic effects, especially in the mountainous areas of the historical maps. The problem and appearance of this error is similar to residuals of the (seven-parameter similarity) transformation parameters between the national coordinate systems and the ETRS89 (i.e. Stopar and Kuhar 2003). These can be suitably represented by an autocorrelated random surface. For the locally systematic error we assumed a spatially smooth distribution (Fig. 7.8a). Locally/regionally shifted homogenous areas were simulated; for example, one valley or one settlement is shifted in context on the map.

Such homogenous areas are difficult to classify, but they may be simulated. The same foundations were used to apply spatially rough effects related to errors on short distances. They can be approximated with more locally thin and random distribution. This error distribution is quite similar to the first principle described above, but instead of lines representing boundaries, small areas were simulated as smooth and rough continuously varying error distribution surfaces (Fig. 7.8).

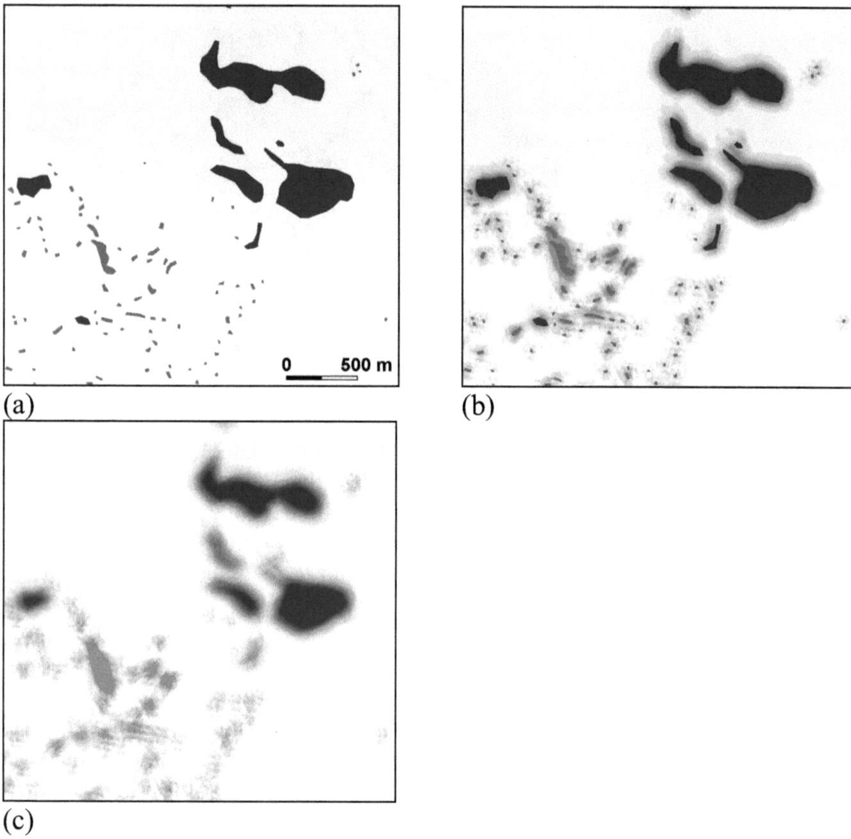

Fig. 7.7. Two selected land uses, buildings (red) and moors (blue), (a) digitised and attributed for the area of Pokljuka due to TK25 is compared with fuzzified boundaries produced from polygons simulated 100 times with the MC method. Parameters were chosen attributing the error model for each category differently to (b) the area around original land uses and (c) and equally weighted simulated polygons. The mixed (confused) fuzzy classified areas in (b) and (c) are slightly greenish. The visualisations use red and blue composites (area of 2.6 by 2.5 km).

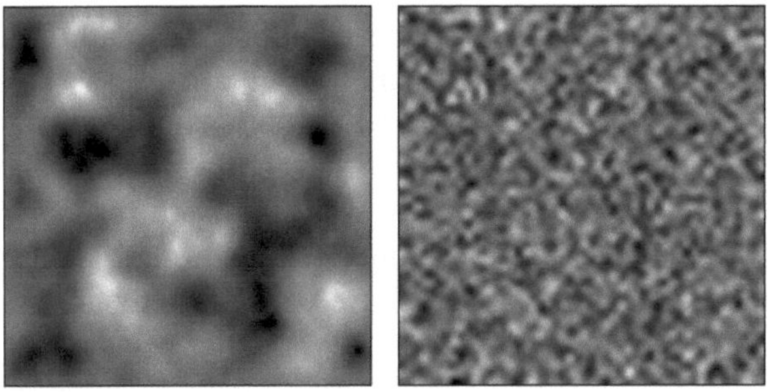

(a) (b)

Fig. 7.8. Random distribution simulated as smooth (a) and rough (b) continuously varying error distribution surfaces

Both artificially autocorrelated random surfaces were generated following a standardised unconditional normal distribution. Moran I was measured to achieve the required degree of autocorrelation (Haining 2003). This procedure was performed for the smooth and rough random surfaces bearing in mind their different spatial resolutions. The procedure is based on the exchanging of randomly selected points on the random surface that covers the study area and autocorrelation is controlled with the Moran I. Exchanged points were accepted at each step if the autocorrelation was higher than in the previous step. This procedure requires many iterations.

After the smooth and rough surfaces were computed (see Fig. 7.8), they were combined to produce a more complex error surface, considering empirical results of discovery and a determination of their portion. The value of the portion comes from the relationship between many imprecise measurements between selected local/regional areas, and has a higher precision inside of those areas. Specifically, the chosen portion of the smooth surface was much higher than that of the rough one, but could be different for each particular map source or for wider regional areas (mountains vs. plains).

As with the first principle (described in the previous section), the approximated σ = 515 m was annotated to the standardised unconditional complex error surface. This required attributing shifts in the x and y directions of the Cartesian coordinate system due to complex error surfaces. The shifts were applied using two approaches. The first was tessellation of the land use areas into uniform square areas (vector polygons) with 200 m sides based on attributes of land use categories. Each tessellated square was shifted independently in the x and y directions according to the values of the complex error surface. This approach was additionally combined with the second approach to the first principle described in the previous section, so the boundaries of all the square polygons were simulated

(Fig. 7.9). Thus we effectively simulated error distribution due to all three presumed effects: random, locally systematic and systematic/gross. This requires a lot of processing time but does not need as many simulations as the first principle. Twelve complex simulated error surfaces were produced for shifts in the x and y directions, thus the error was simulated for N = 6 times. Drecki (2002) represented an attribute uncertainty through square technique that is by some means similar approach to the described one.

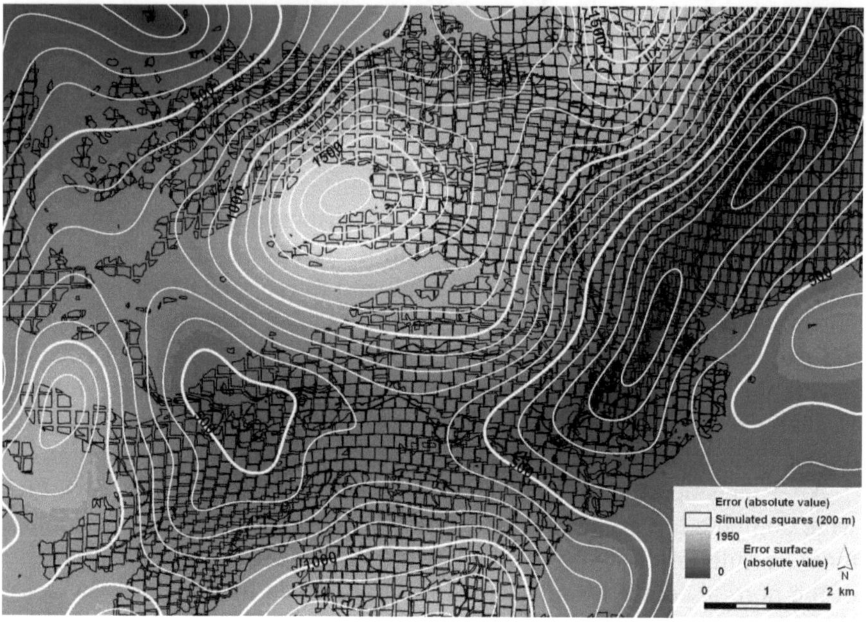

Fig. 7.9. One of the N = 6 MC simulations of selected land uses from the JMT28.8. A random distribution surface represents absolute values between coordinates shifted in the x and y directions. Brighter areas mean larger shifts. Contour lines support shaded surface – higher density lines represent conversion between different locally systematic errors. The autocorrelated arrangement of square polygons with 200 m sides presents possible distortions of land use data due to shifts. Where squares are connected, the local systematic error is constant; elsewhere it is changing. Individual land use areas are therefore continuously deformed and shifted. Little or no random error is expected within the square boundaries. Some parts of the squares are cut off owing to edges of the land use areas.

The second approach to the second principle is similar to the first approach, but instead of squares as vector polygons, denser grid points were produced to simulate a grid-based land use error distribution surface. The simulations were visualised in a variety of ways. To illustrate, Figure 7.10

presents the sequence and Figure 7.11d presents the effective fuzzification of the summed-up simulation fields. Where possible, the particular land use category is presented by contrast of a particular colour.

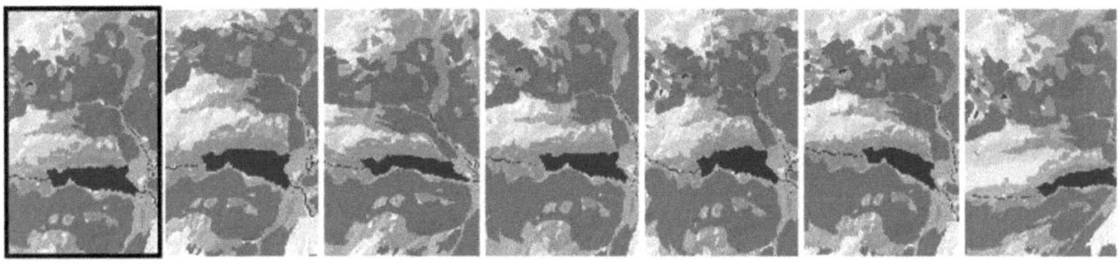

Fig. 7.10. Original land use based on JMT28.8 (in a frame) and N = 6 simulations from the error model (area of 7.5 by 12 km).

The described simulation procedures and visualisations of the both principles (as seen as integrated visualisation in Fig. 7.11) provide better comprehension and cause other error effects to be factored into the error distribution models.

4 Discussion of the Results

In this research an alternative strategy for the evaluation of data quality was introduced. The classical strategy relies on knowledge of original procedures for data acquisition, and consequently, on the expected quality of sources and the final product – maps or other spatial data sets. The availability of a high quality metadata determines the suitability of a data model. For the classical strategy we may be familiar with error distribution models that can vary depending on the procedures for data acquisition and processing: measurements, generalisation, digitisation, attribution (land use categories), etc., or due to different data features (Peterca et al. 1974, Robinson et al. 1995; Khorram et al. 1999). For insight into data and their quality (or even into the data model and its quality), simulations of such error distribution models may not be needed. Standardised statistical tests of actual data quality can be relatively easily applied and evaluated. Nevertheless, those tests can show a distorted understanding of the product without effective visualisation. Additionally, understanding and correcting some effects (errors) that can appear by more thorough investigations and are probably not due to the well defined data model, may also be necessary.

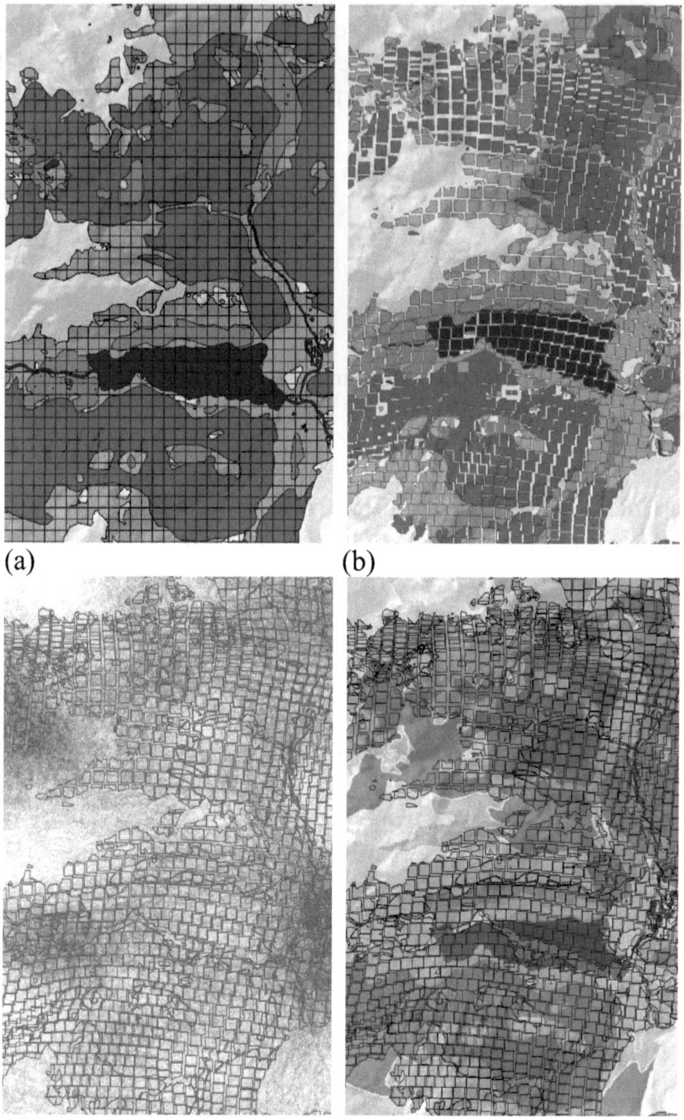

Fig. 7.11. Presentations of JMT28.8 using square polygons with 200 m sides. The example (a) shows squares and land use before simulation, and, after a single simulation; (b) squares with assigned land use; (c) squares in combination with N = 100 simulated lines of the land use boundaries (grey); and (d) squares with the assigned land use in combination with land use before simulation; higher contrast indicates the overlay of both (area of 7.5 by 12 km).

Our alternative strategy relies on acquiring knowledge of the data and their quality and, when necessary, on data models and their quality as well (as data models for older data are not clear). It is a supplement to the classical strategy. Our strategy is suitable for evaluation of very complex data for which we don't have much preliminary knowledge, and for which we also don't have any reliable reference comparison data, for example, the historical maps presented in this study. In such cases we can not easily reconstruct a reliable data model or build the error distribution model. Our proposition was that effective visualisation and iterative processes that include quantitative and qualitative methods can help to solve these problems. Even more, when the classical strategy is not feasible, we can apply as many steps as possible of our iterative strategy (methodology) for any data source. In the following list, steps 1–2 and 6–8 are supplemented to the classical strategy:

1. Acquire knowledge (basic metadata) of the history, surveying and mapping techniques, surveyor group, past physical and human landscape, etc. using different qualitative sources, for example books
2. Define and describe the data models and try to understand their quality (using historical maps)
3. Acquire data (digitised historical maps) with georeferencing or transformation with known parameters, and vectorisation of selected features (land use) (standard routine)
4. Perform statistical quality control (of historical maps) and analysis of techniques (standard routine)
5. Apply alternative quantitative and especially qualitative (visual) methods for an improved understanding of the data (historical maps and acquired land use) (see Wood and Fisher 1993; Burrough and McDonnell 1998; Podobnikar 2005)
6. Propose a hypothesis for the error distribution models based on data quality parameters. The classification of the proposed distributions came from steps 1 and 5, together, rather than from step 4 alone
7. Apply MC methods for simulating the proposed (positional) error distribution model, described statistically
8. Classify and visualise the error by means of fuzzy set theory; obtain a better data and error distribution model
9. Correct georeferencing, if necessary

We propose additional steps towards the spatial analyses of different data sets, which were not described in this paper. These steps may independently extend both the classical and our alternative strategy:

10. Eliminate any more complex systematic errors that are discovered

11. Apply a common (standardised) data model for all data sets and homogenise them (land uses acquired from historical maps of different ages) following from step 2 (see Fig. 7.12)
12. Implement reliable spatial analyses in GIS that may include fuzzy boundaries or fuzzy transition areas for handling (for example land use, derived from different historical maps)

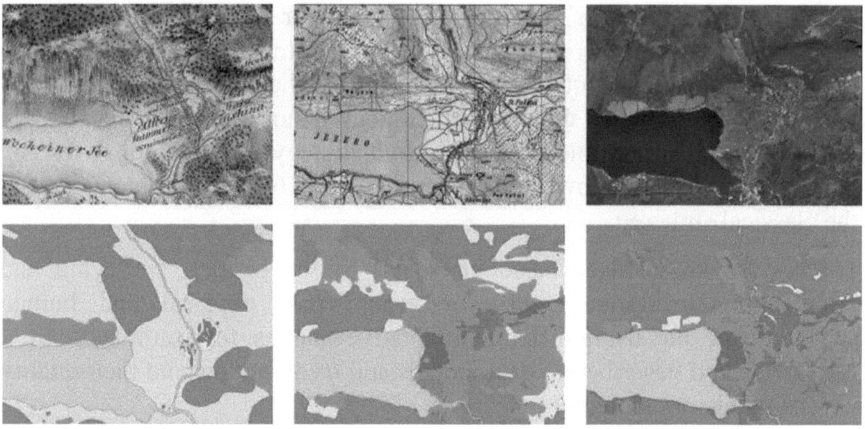

(a) (b) (d)

Fig. 7.12. (a) JMT28.8, (b) TK25 and (c) contemporary orthophoto (upper row) and associated land use data (lower row) presented on area of 4.2 by 3 km. Besides temporal changes of land use, positional accuracy increases significantly with time, considering unique scale. How accurate should a common data model be (nominal ground) for the mutual spatial analysis in GIS?

We now turn to a more detailed discussion of simulation methods. Although other data quality elements were evaluated, we focused on positional ones. For randomisation of the possible and proposed error distributions, powerful MC simulation methods were applied. Three principal types of error distributions were proposed: random, locally systematic, and systematic with the limit at the gross error. They were applied within two main principles: vector boundary simulation and continuous surface simulation. The first principle was applied to the land use data with 100 repetitions, using an appropriate correlation coefficient between neighbour vector nodes, addressing random or systematic error. Then followed a rasterisation, and summing up all the data sets by means of two approaches: lines presented as fuzzy boundaries and polygon areas presented as fuzzy transition zones. Similarly, the second principle, continuous surface simulation, comprised two approaches. Both required attributing the planar shifts to account for the randomly distributed continuously varying error surface that was produced. The shifts were then applied to the land use data and simulated with six repetitions using two approaches: as polygons with square areas having 200 m sides and as grid points that were summed up as fuzzy areas with fuzzy borders. The two principles produced different views of possible positional error distribution models, especially in the areas where we couldn't easily acquire quality parameters. Continuous surface simulation of error distribution, especially, seems to be a very powerful solution. It can be easily combined with other simulated surfaces that present other characteristics or with errors that are inherent in the land use data set. Actually, the error models are only generalisations and minor possible approximations of the real world complexity. For more complex and reliable simulations, more knowledge of the error distribution is needed.

The JMT28.8 maps and derived land use data were analysed in greater detail. Figures 7.6–7.7 and 7.9–7.11 visualise fuzzy borders and transition areas between land use classes that can occur in the unsettled areas that were not precisely surveyed in the 18th century, and consider less objective and precise digitisation of land use from the georeferenced map sheets. Visualisation of fuzzy features and other presentations of simulations were considered to be more illustrative than simple epsilon-bands, and significantly more illustrative than presentations of statistical evaluations by numbers that present quality.

5 Conclusions

Spatial phenomena are not easily transformable to information based on spatial data models in terms of a reliable interaction between spatial positions and attributes in GIS. Different data models may perhaps have been used during data updates in the past. Later, the producers might have semantically understood differently or even incorrectly the topical data models for particular data during their lifetime. This can mask issues of model quality and data quality. The most important output of this research was to recommend that simulation and visualisation methods be included within the proposed alternative strategy to increase the knowledge and understanding of the spatial data and the processes that underlie them, from the first ideas via abstraction for establishing a data model, to realisation of the data.

There are several difficulties with the use of the data sets in this study of historical maps and derived land use data. The more obvious ones are positional and thematic. In addition, temporal accuracy/error might be important, especially when the mapping took place over a longer period. The problem is more evident when we want to use two data sources from nearly the same period but from neighbouring countries to acquire homogeneous land use data. Even with contemporary data an associated problem may occur; owing to the long-standing legacy of measurements of the features, many traces (and errors) were propagated from the older data (maps). Therefore the actual data products are a kind of collage. Using spatial data of varied quality as an input for reliable mutual spatial analysis may be a poor decision if heterogeneity and uncertainties of the data are not successfully resolved. But, unfortunately, describing the quality of such data can be a complex task. Some more types of errors were partly uncovered and resolved within this paper's proposed strategy.

To increase understanding of the data models and their applicability to historical maps, MC simulation methods were applied. The simulations were based on three proposed stochastically-described positional error models. The understanding of many aspects of the data's nature was increased with visualisation of possible error distributions. Effective visualisations seem to be more useful to the experienced operator than just pure numerical outputs. A whole methodology of getting experiences from studied data sets was introduced within an alternative strategy for evaluation of data quality. It includes a "classical" one, but instead of just standard outputs, more efforts were invested into considerably deeper knowledge of data and the processes carried out on them.

Explanation and objectification of more complex but significant phenomena hidden within spatial data is a great challenge for the expansion of this study. We proved that a more effective research method is to study particular problems on the actual applications using real datasets, rather than try to generalise a theory again – applying iterative and so called abductive approaches. A successful study should be supported by other aspects, even by in situ observations, admitting the capabilities of researchers for objective interpretations of the problems. This can result in a distinct improvement in the quality of the data and metadata (knowledge) suitable for spatial studies. Thus the procedures for consistent analyses can be significantly simplified, making the concept clearer and the process more effective. In extreme circumstances, when high quality data and a perfect overview of the situation is available, an equivalent reliable simplified solution may be found.

Acknowledgments

The first steps of the research on the Monte Carlo simulations were supported by the Ministry of Higher Education, Science and Technology of Slovenia. Applied outputs of this research were utilized within the Interreg IIIB, SISTEMaPARC project where our team studied historical maps used for land use dynamics analysis. The final part has been carried out within the project TMIS.plus.II, funded by the Austrian Research Promotion Agency within the ASAP programme. I'm very grateful for all the support from my family and friends, Zoran Stančič, Theo Tijssen, Norbert Pfeifer and other anonymous reviewers for their suggestions.

References

Aalders HJGL (1996) Quality metrics for GIS. In: Advances in GIS Research II: Proceedings 7th International Symposium on Spatial Data Handling, 12–16 August 1996. Delft, The Netherlands. Taylor & Francis, London, pp 5B.1–5B.10

Anselin L (2005) Mapping and Geovisualization. In: ACE 592SA Spatial Analysis. Avalable online at http://sal.uiuc.edu/courses

Bernhardsen T (1992) Geographic information Systems. Norwegian Mapping Authority, Viak IT

Blakemore M (1984) Generalization and error in spatial databases. Cartographica 21:131-139

Box GEP, Muller ME (1958) A Note on the Generation of Random Normal Deviates. Annals of Mathematical Statistics 29:610-611

Brown DG, Duh J-D (2004) Spatial simulation for translating from land use to land cover. International Journal of Geographical Information Science 18:35-60

Brown JD, Heuvelink GBM (2007) The Data Uncertainty Engine (DUE): A software tool for assessing and simulating uncertain environmental variables. Computers & Geosciences 33:172-190

Buckner B (1997) The Nature of Measurement: Part IV-Precision and Accuracy. Professional Surveyor Magazine 17

Burrough P, McDonnell R (1998) Principles of Geographical Information Systems—Spatial Information Systems and Geostatistics. Oxford University Press, Oxford

Clarke KC, Teague PD (1998) Cartographic symbolization of uncertainty. In: ACSM Annual Conference, 2–3 March 1998. Baltimore, MD, pp 312–323

Cressie NAC (1993) Statistics for Spatial Data. John Wiley & Sons, New York

Drecki I (2002) Visualisation of Uncertainty in Geographic Data. In: Shi W, Fisher PF, Goodchild MF (eds) Spatial Data Quality. Taylor & Francis, New York, pp. 140-159

Ehlschlaeger CR, Shortridge AM, Goodchild MF (1997) Visualizing Spatial Data Uncertainty Using Animation. Computers & Geosciences 23:387-395

Fisher PF (1994) Probable and Fuzzy concepts of the uncertain viewshed. In: Worboys M (ed) Innovations in GIS 1. Taylor & Francis, London, pp 161-175

Fisher PF (1996) Animation of reliability in computer-generated dot maps and elevation models. Cartography and Geographic Information Systems 23:196-205

Giordano A, Veregin H (1994) Il controlo di qualità nei sistemi informativi territoriali (A quality control in GIS, in Italian). il Cardo, Venezia

Hengl T, Walvoort DJJ, Brown A (2002) Pixel and Colour Mixture: GIS Techniques for Visualisation of Fuzzyness and Uncertainty of Natural Resource Inventories. In: Hunter G (ed) Proceedings International Symposium on Spatial Accuracy in Natural Resources and Environmental Science. Melbourne

Haining R (2003) Spatial Data Analysis: Theory and Practice. Cambridge University Press

Heuvelink GBM (1998) Error Propagation in Environmental Modelling. CRC Press

Kanevski M, Maignan M (2004) Analysis and Modelling of Spatial Environmental Data. EPFL Press

Khorram S, Biging GS, Chrisman NR, Colby DR, Congalton RG, Dobson JE, Ferguson RL, Goodchild MF, Jensen JR, Mace TH (1999) Accuracy Assessment of Remote Sensing-Derived Change Detection. ASPRS, Maryland

Martinoni D, Bernhard LA (1998) Conceptual Framework for Reliable Digital Terrain Modelling. In: Proceedings of the 8th International Symposium on Spatial Data Handling '98, 11–15 July 1998. Vancouver, BC, Canada, pp 737–750

Molenaar M (1998) An Introduction to the Theory of Spatial Object Modelling for GIS. Taylor & Francis, London

Openshaw S, Charlton M, Carver S (1991) Error propagation: a Monte Carlo simulation. In: Masser I and Blackmore M (eds) Handling Geographical Information: Methodology and Potential Applications. Longman, London, pp 78–101

Perkal J (1956) On the ε-Length. Bulletin de l'Académie Polonaise des Sciences 4/7: 399-403

Peterca M, Radošević N, Milosavljević S, Racetin F (1974) Kartografija (Cartography, in Serbo-Croatian). Vojnogeografski institut, Beograd

Plewe B (2002) The Nature of Uncertainty in Historical Geographic information. Transactions in GIS 6:431–456

Podobnikar T (2005) Production of integrated digital terrain model from multiple datasets of different quality. International Journal of Geographical Information Science 19:69-89

Podobnikar T (2009) Geocoding and quality assessment of Josephine survey maps for the mountainous region in the Triglav National Park. Acta Geodaetica et Geophysica Hungarica 44 (forthcoming)

Rajšp V, Ficko M (1995) Slovenija na vojaškem zemljevidu 1763–1787 (Josephinische Landesaufnahme 1763–1787 für das Gebiet der Republik Slowenien) (Slovenia on a military map, in Slovenian) 1. ZRC SAZU, Ljubljana

Robinson AH, Morrison JL, Muehrcke PC, Kimerling AJ, Guptill SC (1995) Elements of Cartography. John Wiley & Sons, New York

Shi WZ, Ehlers M, Tempfli K (1999) Analytical Modelling of Positional and Thematic Uncertainties in the Integration of Remote Sensing and Geographical Information Systems. Transactions in GIS 3/2:119-136

Stopar B, Kuhar M (2003) A study of distorsions of the primary triangulation network of Slovenia. Acta Geodaetica et Geophysica Hungarica 38:43-52

Taylor BN, Kuyatt CE (1994) Guidelines for Evaluating and Expressing the Uncertainty of NIST Measurement Results, NIST Technical Note 1297. U.S. Government Printing Office, Washington

Wood JD, Fisher PF (1993) Assessing interpolation accuracy in elevation models. IEEE Computer Graphics and Applications 13:48-56

CHAPTER 8

Global Morphometric Maps of Mars, Venus, and the Moon

Igor V. Florinsky

Institute of Mathematical Problems of Biology, Russian Academy of Sciences, Pushchino, Moscow Region, Russia

Abstract

The paper examines capabilities of digital terrain modelling to analyse topography of Mars, Venus, and the Moon at the global scale. Digital elevation models with a resolution of 30 arc-minutes were used as initial data. Fifteen local, regional, and combined topographic variables were for the first time calculated and mapped for the entire surface of the celestial bodies. In terms of geomorphological and geological interpretation, the most useful were maps of horizontal and vertical curvatures as well as catchment and dispersive areas. On catchment area maps of Mars and Venus, it was possible to detect several helical structures encircling the planets from pole to pole. Their origin may be connected with palaeorotational planetary stresses. A map of topographic index may be used to study the spatial distribution of the water ice deposits in the Martian soil.

Keywords: digital terrain modelling, surface, geology, planet

1 Introduction

Topography, resulting from endogenous and exogenous geophysical processes of various spatial and temporal scales, carries information both on these processes and target properties. This has led geoscientists to use morphometric analysis at different scales. At present, digital terrain modelling is the basis for investigations of this kind (Moore et al. 1991; Florinsky 1998a; Wilson and Gallant 2000).

Digital terrain models (DTMs) are widely used in research of the planets of the Earth group, satellites, and asteroids, including refinements of the shape of a celestial body (Sadovnichy et al. 1991; Smith et al. 1997; Nyrtsov 2000), studies of the global and regional tectonic structures (Solomon et al. 1991; Zuber et al. 1994; Watters et al. 2001; Nimmo et al. 2003), prediction of the crust structure (Zuber et al. 2000; Wieczorek 2007), modelling of lithospheric evolution (Smith et al. 1999; Byrne 2007) and mantle convection (Simons et al. 1994), statistical characterisation of the relief (Lucey et al. 1994; Barnouin-Jha et al. 2008), geomorphic mapping (Bue and Stepinski 2006), studies of volcanic (Neukum et al. 2004), glaciological (Fishbaugh and Head 2001; Neukum et al. 2004), and erosional processes (Williams et al. 2005), modelling of impact structures (Cochrane and Ghail 2006), reconstruction of palaeohydrological events (Carr and Head 2003) and palaeoclimatic conditions (Stepinski and Stepinski 2005), valley network detection (Jenson 1991; Molloy and Stepinski 2007), and selection of landing sites (Kirk et al. 2003).

Local topographic variables of the complete system of curvatures (Shary 1995) and regional topographic variables (Speight 1974) are key parameters of geomorphometry. However, they have not been used in planetary studies at the global scale. A recent analysis of the Earth's global topography demonstrated the usefulness of these attributes (Florinsky 2008). This paper examines capabilities of digital terrain modelling to analyse the global topography of Mars, Venus, and the Moon.

2 Materials and Methods

The study was based on 30 arc-minute gridded global digital elevation models (DEMs) of Mars, Venus, and the Moon. A DEM of Mars (Fig. 8.1a) was obtained from the Mars Orbiter Laser Altimeter Mission Experiment Gridded Data Record (MOLA MEGDR) archive (Smith 2003) of the Mars Global Surveyor mission (Albee 2000). A DEM of Venus (Fig. 8.1b) was compiled using data from two archives of the Magellan mission

(Saunders and Pettengill 1991), viz. Global Topography Data Record (GTDR) (Ford 1992) and Magellan Spherical Harmonic Models and Digital Maps (MSHMDM) (Sjogren 1997). A DEM of the Moon (Fig. 8.1c) was extracted from the Clementine Gravity and Topography Data (CGTD) archive (Zuber 1996) of the Clementine mission (Nozette et al. 1994).

Each DEM included 260,281 points (721 by 361 matrices). Martian and Lunar elevations were presented with reference to related geoids (Zuber et al. 1994; Smith et al. 1999), while Venusian ones – with reference to the mean planetary radius (Rappaport et al. 1999).

Any DEM includes high frequency noise leading to the derivation of useless, noisy digital models and unreadable maps of topographic variables (Florinsky 2002). The problem can be partially resolved by DEM smoothing. One, two, and three smoothing iterations were applied to the DEMs using a 3 by 3 kernel with linear inverse distance weights.

For Mars, Venus, and the Moon, digital models of fifteen topographic attributes were derived: twelve local variables (i.e., horizontal, vertical, accumulation, difference, ring, minimal, maximal, mean, Gaussian, unsphericity, horizontal excess, and vertical excess curvatures), two regional variables (i.e., catchment and dispersive areas), and a combined variable, topographic index (*TI*). Definitions, formulae, and interpretations of the variables can be found elsewhere (Florinsky 1998a; Shary et al. 2002).

The method designed for a spheroidal trapezoidal grid (Florinsky 1998b) was applied to calculate local variables. A spheroidal trapezoidal moving window was used in calculations. Its linear sizes depend on the latitude. Known formulae for the inverse geodetic problem (Vincenty 1975; Morozov 1979, pp 178–179) were employed to estimate window sizes during calculations (Florinsky 1998b, 2008). These formulae were also applied to estimate window sizes and point weights during DEM smoothing.

Calculation of regional variables from a DEM is based on logistic procedures, such as cell-to-cell flow line routing. These variables were derived by a single flow line direction algorithm with preliminary filling of sinks (Martz and de Jong 1988) adapted for a spheroidal trapezoidal grid. In a given pixel, a catchment area equals a total area of pixels passed by flow lines arriving the given pixel. An elementary spheroidal trapezoidal area depends on the latitude; it can be estimated by the known formula (Morozov 1979, p 34).

TI is a function of slope steepness and catchment area, that is, local and regional variable, correspondingly. Hence, for an spheroidal trapezoidal DEM, *TI* calculation is based on both of approaches described above.

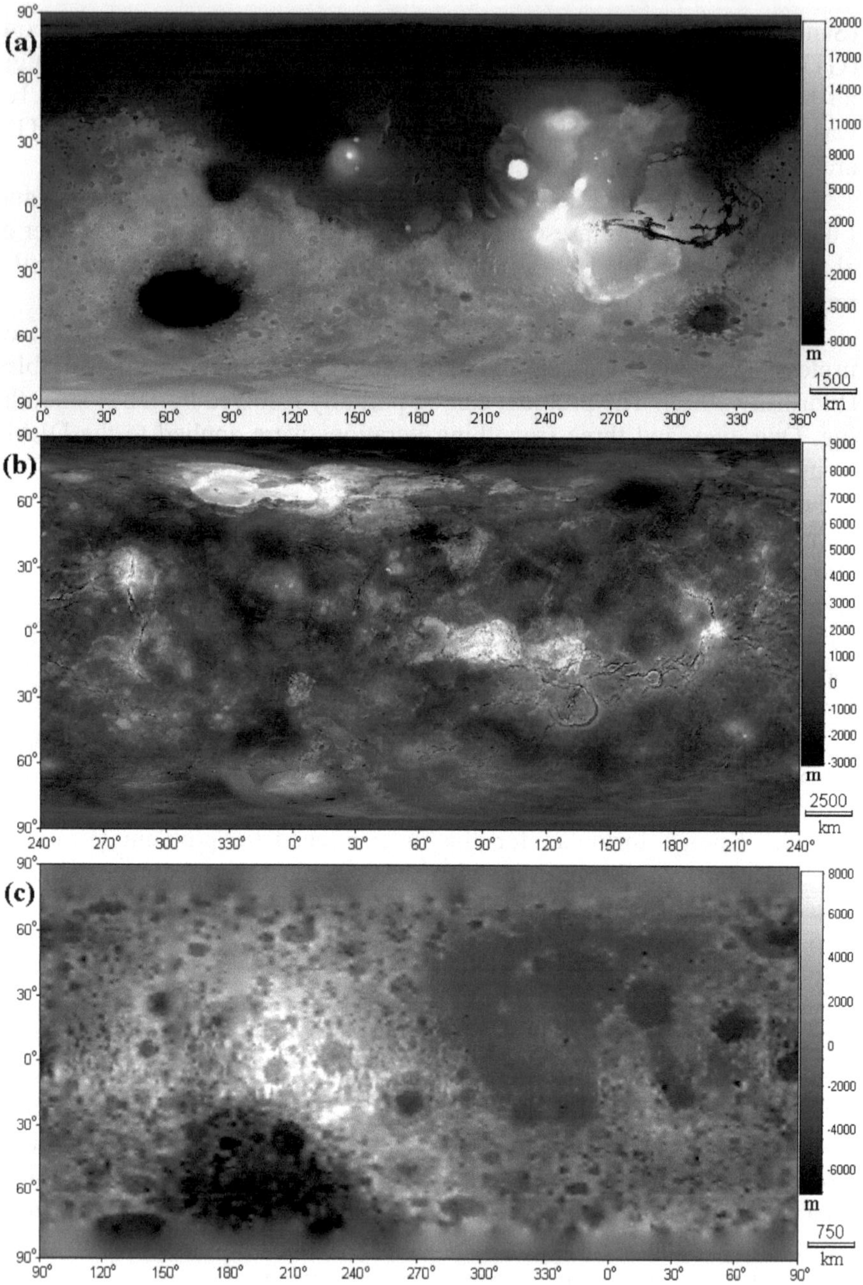

Fig. 8.1. Elevation: (a) Mars, (b) Venus, and (c) the Moon

To estimate linear sizes of an spheroidal trapezoidal window, standard values of the major and minor semiaxes of the Martian ellipsoid were used: 3,396,190 and 3,376,200 m, correspondingly (Seidelmann et al., 2002). Venus and the Moon were considered as spheres of 6,051,848 and 1,738,000 m radii, correspondingly. The DEMs were processed as virtually closed spheroidal matrices. DTMs were derived from smoothed DEMs. DTMs had the resolution of 30 arc-minutes consisting of 721 columns by 361 rows.

It is undesirable to map a topographic variable with an equal-step quantification of its values. This usually leads to information loss due to the large dynamic range of a digital model. To gain a better representation and understanding of global morphometric maps, DTMs were logarithmically transformed as follows (Eq. 8.1):

$$\Theta' = \text{sign}(\Theta) \cdot \ln(1 + 10^n |\Theta|), \tag{8.1}$$

where Θ represents a topographic variable, $n = 0$ for regional variables, $n = 16$ for accumulation, ring, and Gaussian curvatures, and $n = 9$ for other attributes (Figs. 8.2–8.5)

This sort of transformation (Shary et al. 2002) allows one to hold ranges of positive and negative values of a variable (Figs. 8.2–8.3). *TI* was not transformed as its formula includes logarithm calculation. Catchment and dispersive areas were also mapped classifying their values into two levels (Fig. 8.7). The Plate Carrée projection was used to map all variables (Figs. 8.1–8.7). DTM treatment was carried out with LandLord 4.0 (Florinsky et al. 1995).

3 Results and Discussion

3.1 General Geomorphometric Interpretation

Global maps of topographic variables represent peculiarities of the megarelief of Mars, Venus, and the Moon in different ways, according to the physical and mathematical sense of a particular variable.

Horizontal curvature (Fig. 8.2) delineates areas of divergence and convergence of slope lines (positive and negative values, respectively) (Florinsky 1998a; Shary et al. 2002). These areas correspond to valley and ridge spurs (dark and light patterns, respectively), which form so-called flow structures.

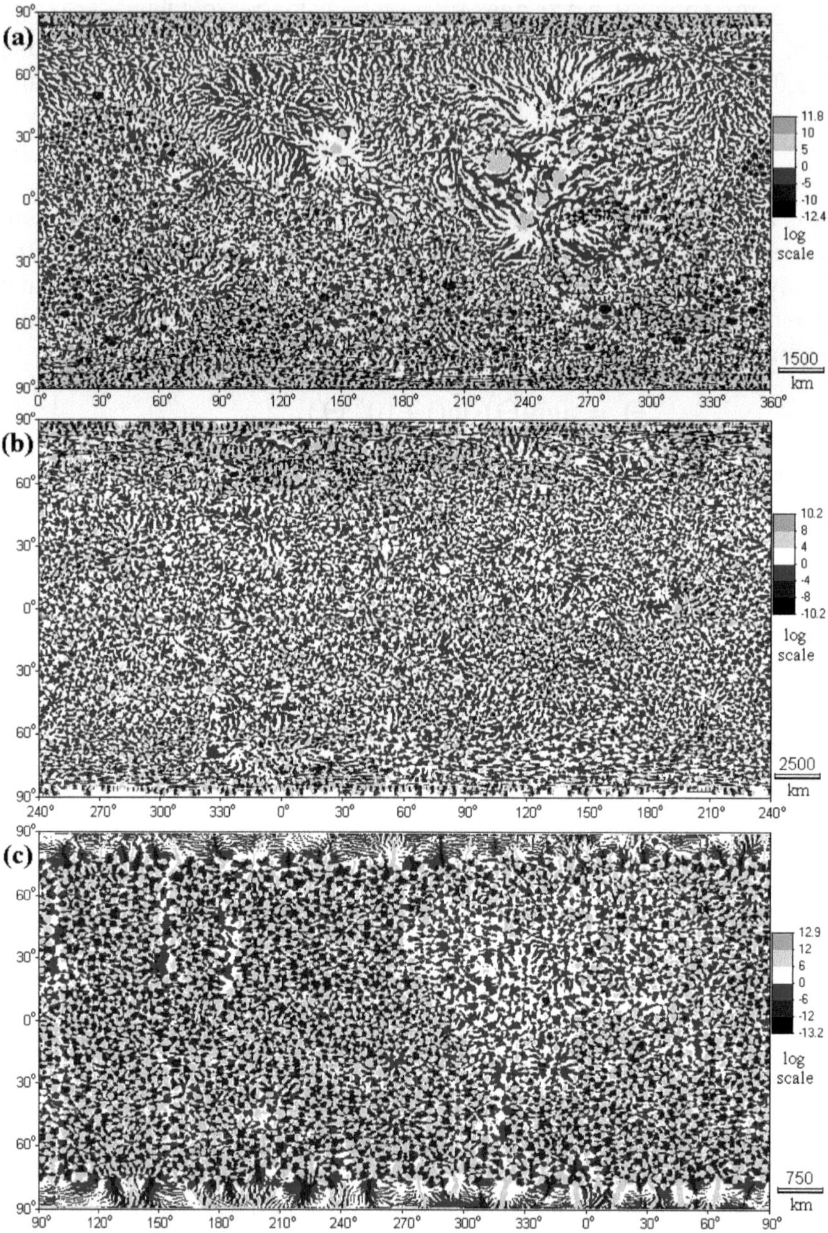

Fig. 8.2. Horizontal curvature derived from the 2-times smoothed DEMs: (a) Mars, (b) Venus, and (c) the Moon

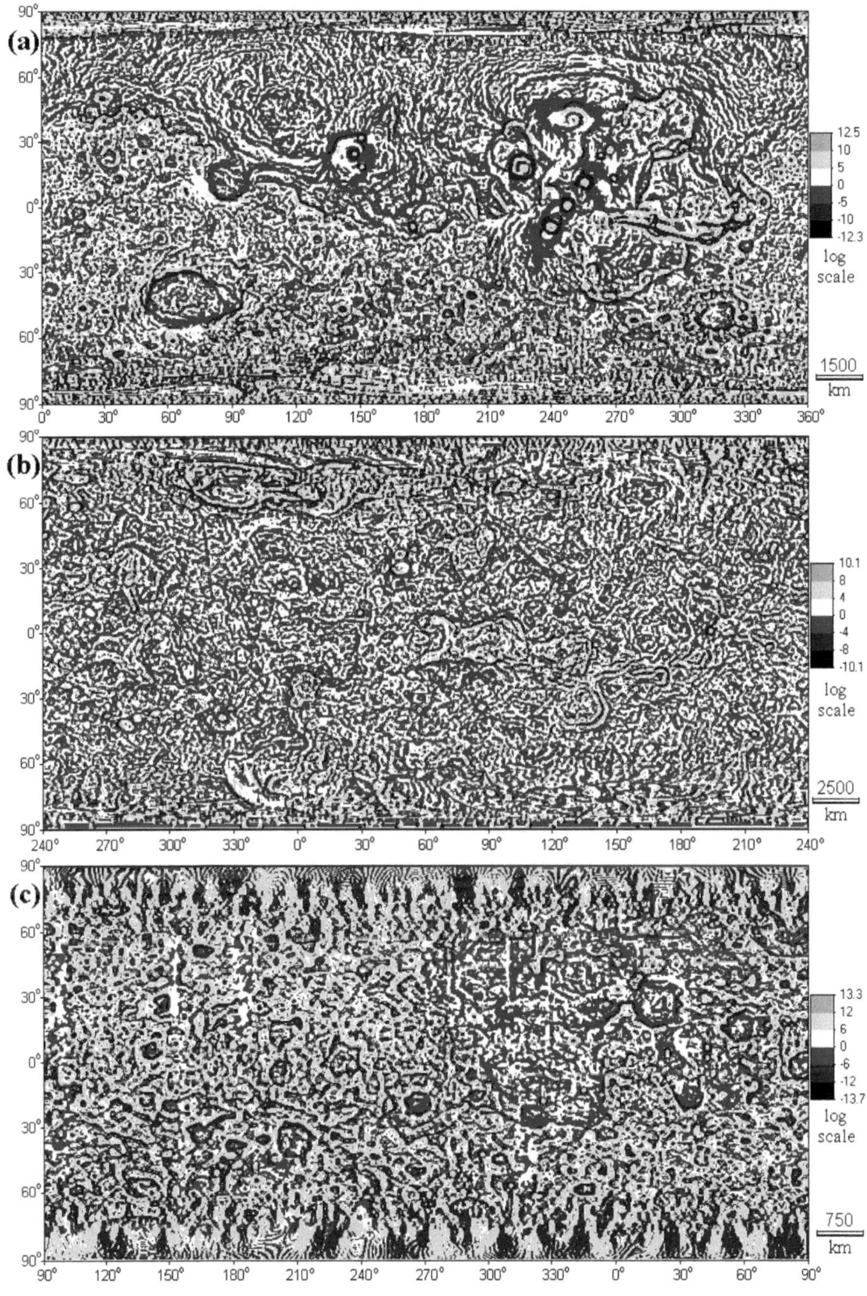

Fig. 8.3. Vertical curvature derived from the 2-times smoothed DEMs: (a) Mars, (b) Venus, and (c) the Moon

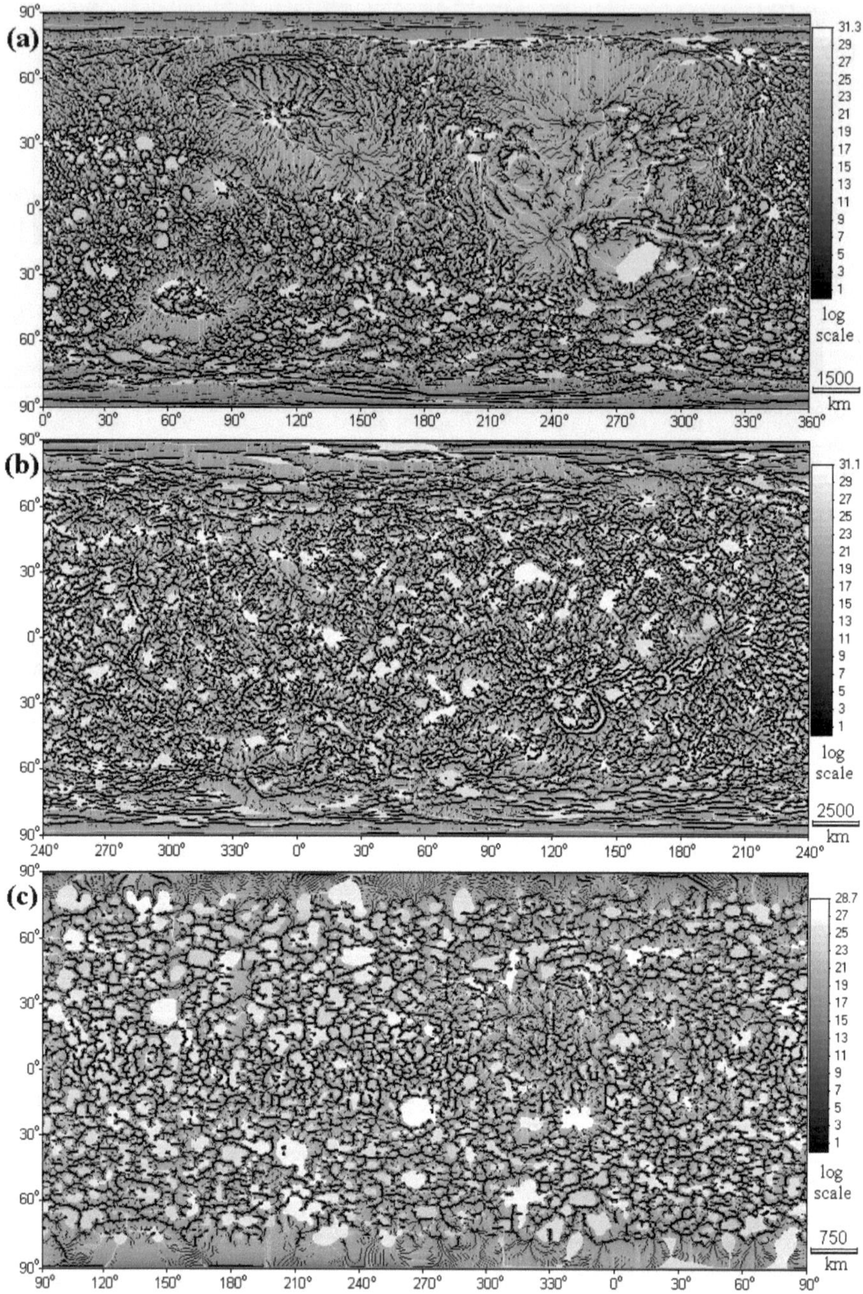

Fig. 8.4. Catchment area derived from the 1-time smoothed DEMs: (a) Mars, (b) Venus, and (c) the Moon

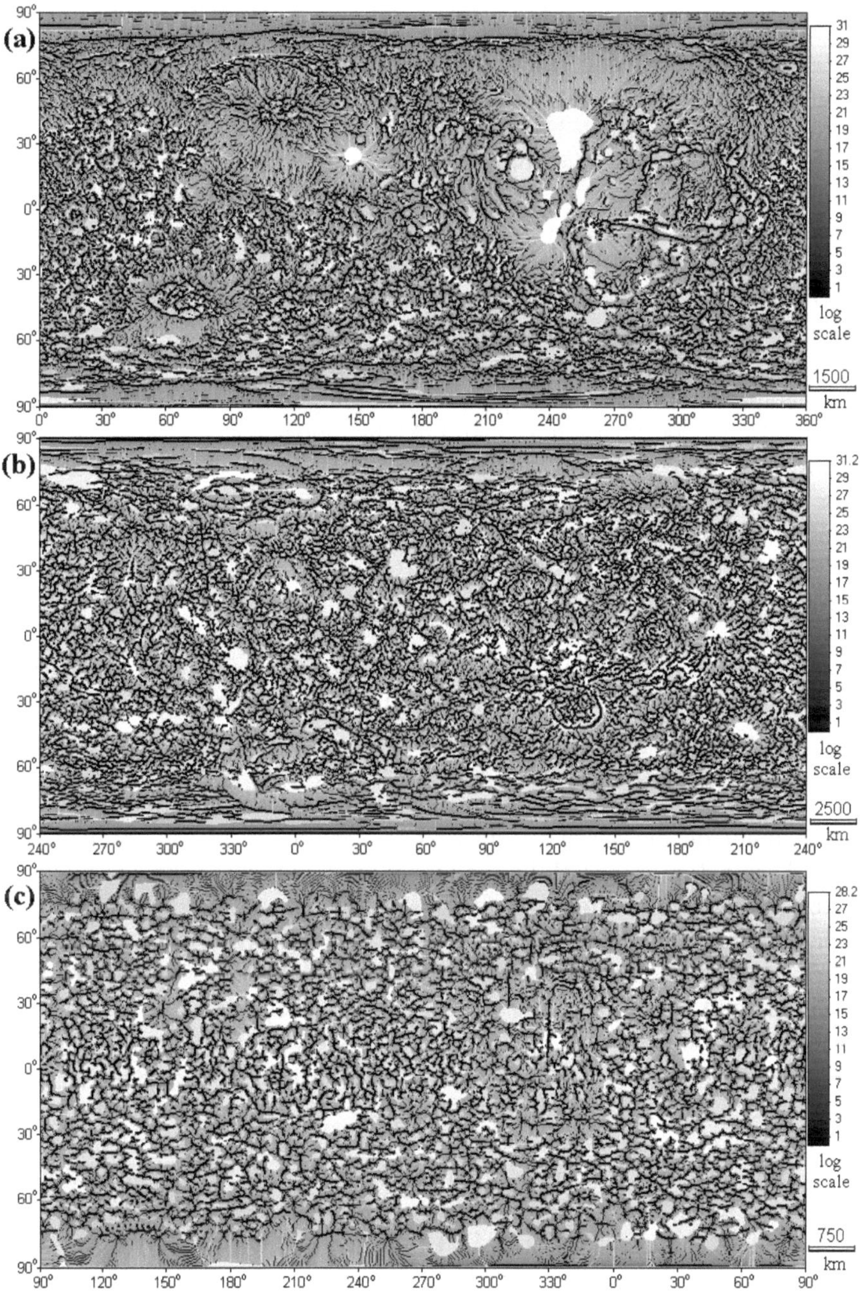

Fig. 8.5. Dispersive area derived from the 1-time smoothed DEMs: (a) Mars, (b) Venus, and (c) the Moon

At the global scale, flow structures are most pronounced on Mars (Fig. 8.2a). In particular, one can see flow structures, probably, of lava origin, beginning on slopes of Alba Patera and forming a huge fan in North Polar Basin (30°–75°N, 200°–310°E). There is a system of flow structures incoming to Utopia Planitia from Nilosyrtis and Protonilus Mensae and Elysium Planitia and Mons (5°–70°N, 75°–150°E). On Venus, flow structures appear only slightly at the global scale (Fig. 8.2b). One can see them on the slopes of Beta Regio (15°–45°N, 270°–300°E). For the Moon, the horizontal curvature represents cellular-like patterns (Fig. 8.2c) resulted from a predominance of craters at the global scale.

Vertical curvature (Fig. 8.3) is a measure of relative acceleration and deceleration of slope lines (positive and negative values, respectively) (Florinsky 1998a; Shary et al. 2002). At the global scale, vertical curvature maps show 'mega-scarps', such as edges of plains, basins, and mountains. For example, on the vertical curvature map of Mars (Fig. 8.3a), one can see borders of Hellas Planitia (30°–50°S, 50°–90°E), Isidis Planitia (10°–25°N, 75°–100°E), Valles Marineris (10°–20°S, 270°–335°E), foothills of Olympus Mons (15°–20°N, 220°–230°E), Alba Patera (30°–50°N, 225°–265°E), etc. Artemis Chasma (30°–45°S, 120°–145°E) and some other surface features are pronounced on the vertical curvature map of Venus (Fig. 8.3b). On the vertical curvature map of the Moon (Fig. 8.3c), one can see well-marked borders of Mare Serenitatis (15°–40°N, 10°–30°E), Mare Crisium (10°–20°N, 50°–70°E), etc.

Catchment area measures an upslope area potentially drained through a given point on the surface (Florinsky 1998a; Shary et al. 2002). At the global scale, low values of the catchment area (Fig. 8.4) delineate ridges as black lines, while its high values show valleys as white lines and depressions as light areas. On the catchment area map of Mars (Fig. 8.4a), one can see the planetary network of valleys and canyons. A large feature, Solis Planum (15°–30°S, 270°–290°E), and a plethora of smaller depressions, predominantly craters, stand out. On the catchment area map of the Moon (Fig. 8.4c), one can see borders of Mare Orientalis (15°–25°S, 255°–275°E), Mare Nubium (20°–25°S, 340°–350°E), etc.

Dispersive area measures a downslope area potentially exposed by slope lines passing through a given point on the surface (Shary et al. 2002). At the global scale, high values of the dispersive area delineate mountains and highlands as light areas (Fig. 8.5). For example, the planetary network of ridges can be observed on the dispersive area map of Mars (Fig. 8.5a). Alba Patera (30°–50°N, 230°–260°E), Tharsis Montes (15°N–15°S, 230°–260°E), and other surface features stand out.

On all maps, recognisable artefacts are typical for polar regions (Figs. 8.2–8.7). They were caused by a relatively low accuracy of description of these areas in the databases used (MOLA MEGDR, GTDR, MSHMDM, and CGTD). Besides, CGTD systematic errors led to several meridian artefacts, such as stripes up to 5° wide (80°S–80°N, 155°E; 10°–60°N, 185°E; etc.), on Lunar maps of elevation, horizontal and vertical curvatures, and dispersive area (Figs. 8.1c, 8.2c, 8.3c, and 8.5c). The Venusian DEM was compiled combining data from GTDR and MSHMDM. These archives are marked by distinct resolution and accuracy. As a result, one can see several linear artefacts – marks of this combination (e.g., 15°–55°N, 325°E; 60°–80°S, 325°–355°E) on Venusian maps of elevation, vertical curvature, and catchment and dispersive areas (Figs. 8.1b, 8.3b, 8.4b, and 8.5b).

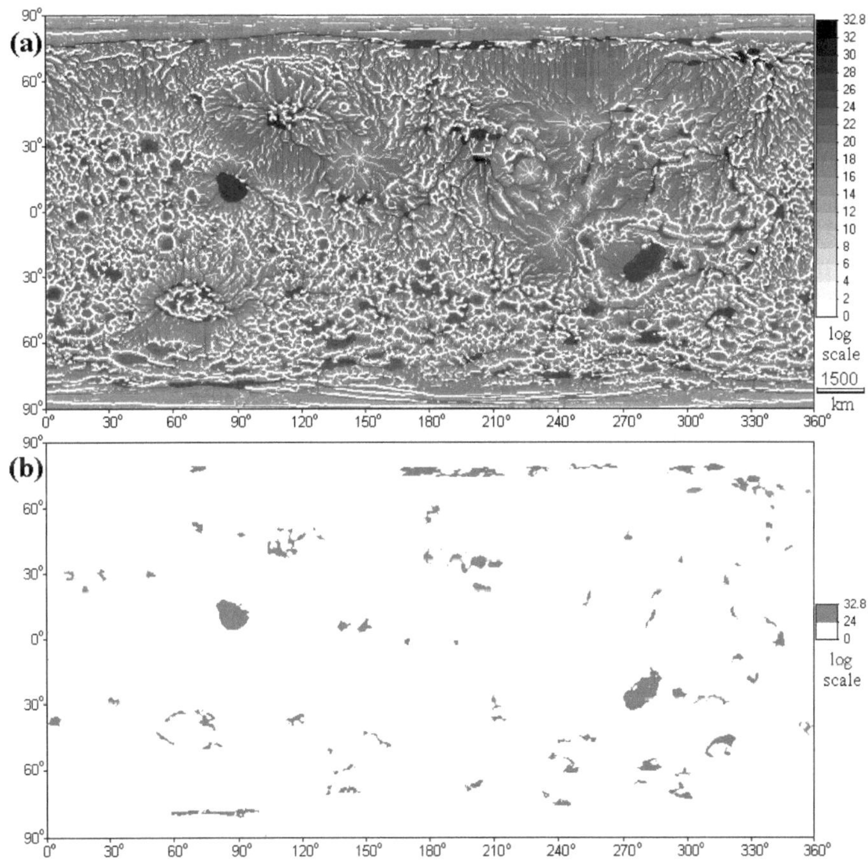

Fig. 8.6. Mars, topographic index: (a) general map, and (b) *TI* values quantified into two levels

There are also computational artefacts on maps of catchment and dispersive areas manifested as straight parallel lines predominantly in polar regions (Figs. 8.4 and 8.5). They are well known artefacts of single flow line direction algorithms common for flat slopes (e.g., Liang and Mackay 2000). However, the artefacts did not influence the perception of topographic features since the artefacts are situated within limited zones of the maps.

3.2 Spatial Distribution of the Water Ice on Mars

One of the topical problems of Mars studies is the elucidation of the spatial distribution of the subsurface water ice (Baker 2001; Cabrol and Grin 2005; Kuzmin 2005; Mitrofanov 2005; Bandfield 2007). It is supposed that water ice (e.g., frozen groundwater, frozen and dust-covered lakes) may be located in the soil to a depth of 1–2 m (Boynton et al. 2002; Mitrofanov 2005; Bandfield 2007). There are several global models for the spatial distribution of the subsurface water ice, particularly based on the neutron spectrometer data (Mitrofanov 2005). It was noted that the Martian topography might theoretically control the spatial distribution of liquid water on the surface: atmospheric pressure exceeds a value critical for the existence of liquid water in some deep depressions (Tokano 2005).

In hydrological studies, *TI* is used to model and predict the spatial distribution of soil moisture and depth to the saturation zone (Beven and Kirkby 1979; Quinn et al. 1995). High values of *TI* are typical for flat territories draining large upslope areas. The higher *TI* value, the higher a soil moisture content, and the less a depth to the saturation zone.

Is it possible to use the global *TI* map of Mars (Fig. 8.6a) to predict the spatial distribution of subsurface water ice? To do this would require, at least, three assumptions. First, at the global scale, relief has varied only slightly after the finish of active tectonic processes responsible for the latest large hydrological events on Mars, such as glacier melting and catastrophic flooding (Baker 2001). Relief changes due to exogenous (e.g., erosion and aeolian) processes can be ignored at this scale. Second, soil water, groundwaters, and lakes turned from liquid into solid state on completion of active tectonic processes and, since that time, ice was preserved intact. Three, there is no ground ice derived from the atmosphere.

If one may accept these assumptions, than *TI* map (Fig. 8.6a) may display the spatial distribution of subsurface water ice. The higher values of *TI* (the darker patterns), the higher an ice content in the soil and the less a depth of 'the ice table'. Quantifying *TI* values into two levels (Fig. 8.6b), one can delineate regions where relief properties have formed prerequisites for the largest deposits of the subsurface ice. Among these are Solis Planum (15°–30°S, 270°–290°E), Isidis Planitia (5°–20°N, 75°–100°E), several zones at Amazonis Planitia (20°–40°N, 175°–215°E) and Utopia Planitia (40°–50°N, 105°–130°E).

It should be stressed that patterns of the *TI*-based prediction of the ground ice distribution differ essentially from those based on other data and models (Boynton et al. 2002; Kuzmin 2005). A further research (e.g., a comparative analysis of remotely sensed (Bandfield 2007) and *TI* data), should demonstrate if *TI* mapping was selected adequately to predict the spatial distribution of subsurface water ice on Mars.

3.3 Global Helical Structures

Classification of catchment area values into two levels allows one to display ridge networks of Mars (Fig. 8.7a), Venus (Fig. 8.7b), and the Moon (Fig. 8.7c). Analysing of the similar binary maps of the Earth, the author revealed five systems of double helical-like tectonic structures encircling the planet from pole to pole (Fig. 8.8). The structures are topographically expressed by patterns of the global ridge network. They are apparently associated with traces of palaeorotational planetary stresses: two double helices are in reasonable agreement with theoretically predicted traces of shear fractures, while another two double helices are in reasonable agreement with ideal traces of cleavage cracks. Various geological phenomena, such as fracturing, faults, as well as crystal and ore deposits are observed along the helical structures (Florinsky 2008).

Question arises: are global helical structures unique for the Earth, or similar features may be observed on other planets of the Earth group and satellites? The catchment area maps of the celestial bodies (Fig. 8.7) were examined in detail. Attention was paid to lineaments running over the entire globe or a hemisphere. Contrary to regional and continental lineaments, global ones are not manifested as uninterrupted linear patterns or their sequences. A global lineament may be visually detected due to traits of the image texture strung out along a line of some direction (Florinsky 2008).

The visual analysis allowed the author to delineate several global lineaments on Mars and Venus (Fig. 8.9). The global helical structures encircle the planets from pole to pole. The structures revealed are obviously helical zones rather than simply lines. Each helical zone transgresses regions dissimilar in respect to their geomorphic and geological composition.

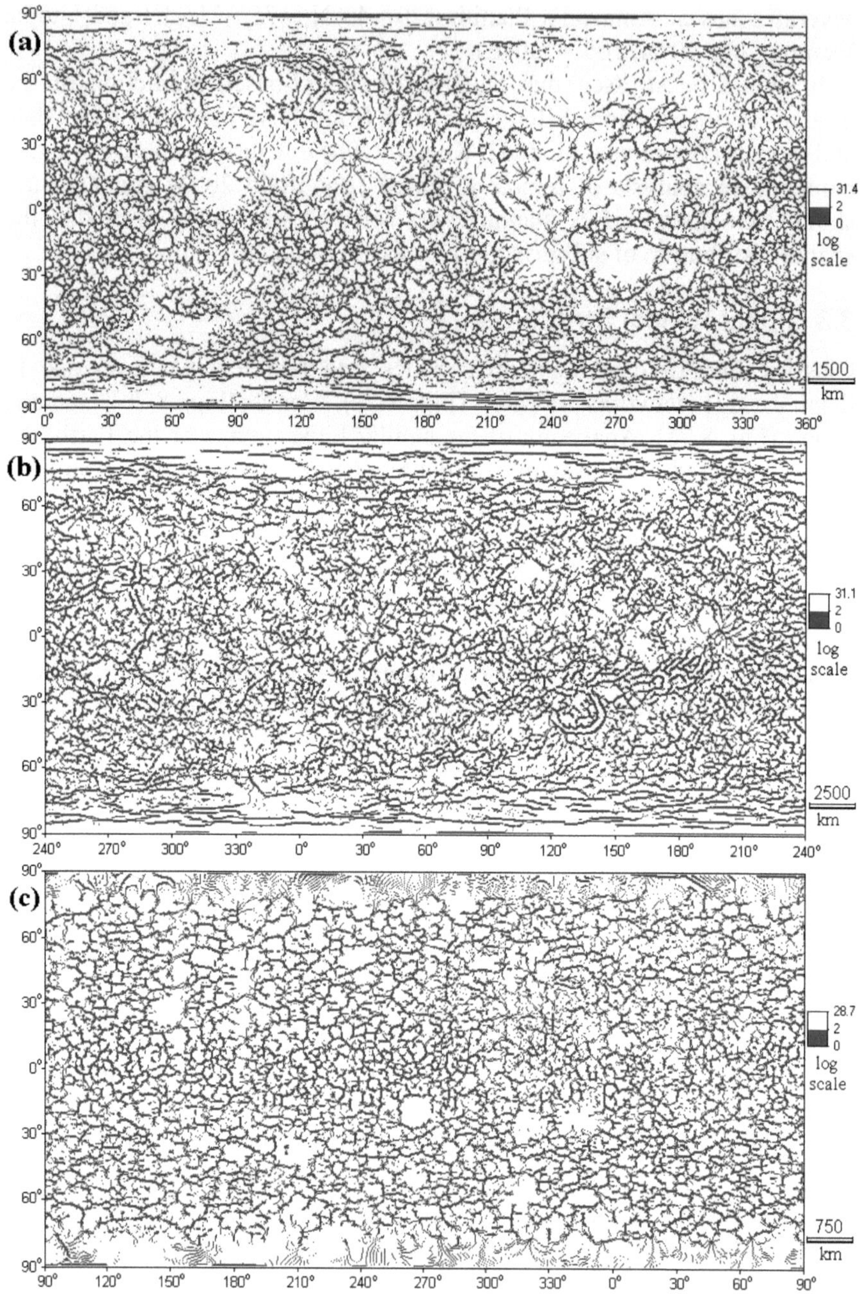

Fig. 8.7. Binary maps of the catchment area: (a) Mars, (b) Venus, and (c) the Moon. Maps of Mars and Venus were derived from the 2-times smoothed DEMs; the Moon map was derived from the 1-time smoothed DEM

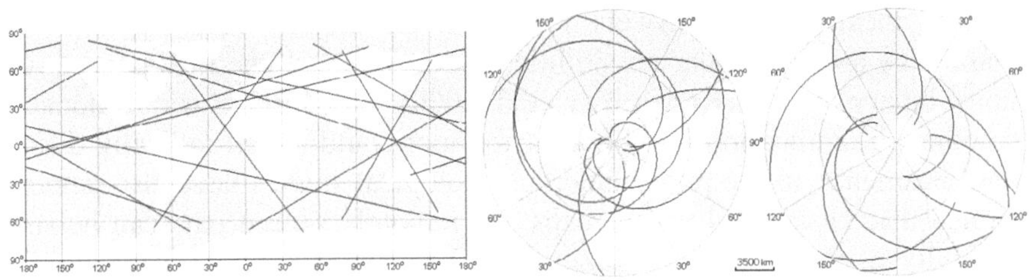

Fig. 8.8. Global helical structures of the Earth; the Plate Carrée projection (left), and polar stereographic projections for the Northern (centre) and Southern (right) hemispheres (Florinsky 2008)

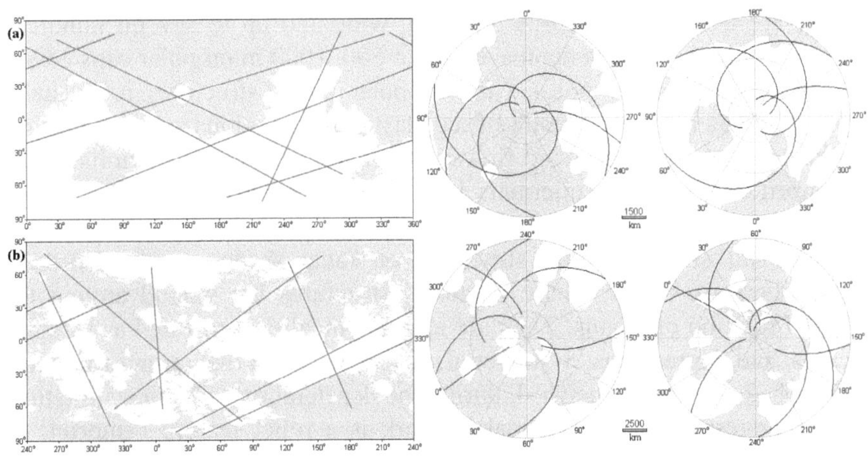

Fig. 8.9. Global helical structures: (a) Mars, and (b) Venus; the Plate Carrée projection (left), and polar stereographic projections for the Northern (centre) and Southern (right) hemispheres. Grey colour indicates areas located below datum

The global lineaments revealed cannot be artefacts due to DEM errors, DEM treatment, or DEM grid geometry (Florinsky 2005). First, noise and errors usually have a random distribution in DEMs. Second, smoothing and derivation of topographic variables were carried out using local filters (n by n moving windows). Third, the grid geometry may amplify its own preferential directions: orthogonal (north-south, east-west) and diagonal (northeast-southwest, northwest-southeast). However, the structures detected have (a) the global character relative to the DEMs, and (b) directions distinct from orthogonal and diagonal ones (Fig. 8.9).

Thus, possible artefacts may be caused by a subjectivity of the visual analysis only. One may speculate that some automated tools (Fukue et al. 1981; Takahashi 1981; Zlatopolsky 1992) should be used to reveal lineaments rather than the visual analysis. However, experiments demonstrated that an ideal observer (Swets 1961) and the human visual system are marked by almost similar possibilities to recognise geometrical patterns on noisy binary images (Krasilnikov et al. 2000). Therefore, it is questionable whether an application of an automated tool could reveal radically different lineaments. In the past, numerous geological studies have also demonstrated that visual analysis of maps and remotely sensed data can be successfully used to detect lineaments (Hobbs 1904; Lattman 1958; Shults 1970; O'Leary et al 1976; Makarov 1981; Trifonov et al 1983).

There is circumstantial evidence that global helical structures (Fig. 8.9) are not artefacts. First, some spiral structures have been documented for the Martian polar caps (Howard 1978, 2000; Fishbaugh and Head 2001). These are troughs and scarps 5–30 km wide and up to several hundreds kilometres long. Trough depths are up to 500–1000 m on polar caps edges, and 100–200 m near poles. Martian polar spiral structures are usually considered as an overall result of ice elastic deformations and sublimation influenced by katabatic wind flows, which are affected by Coriolis forces (Howard 1978, 2000; Weijermars 1985/86; Fisher 1993).

Second, Slyuta et al. (1989) discovered a dense, regular network of dextral and sinistral spiral structures on radar scenes of the Northern hemisphere of Venus. These planetary structures are wound around the axis of rotation of Venus. They are topographically manifested as troughs, scarps, and depressions. Slyuta et al. (1989) believed that strong rotational forces had formed the network during the deceleration of Venus's rotation. They suggested that the helical network is a relict feature, 'imprint' of ancient rotational stress fields, because the current rotation velocity of Venus is quite slow. The low intensity of erosion has allowed relict helical structures to persist on the Venusian surface.

The author supposes that origin of global helical structures of Mars and Venus (Fig. 8.9) may be connected with palaeorotational planetary stresses. Detailed analysis of global helices of Mars and Venus is the objective of further studies.

It is notable that global lineaments were not found on the catchment area map of the Moon (Fig. 8.7c). This may be connected with peculiarities of the rotational regime of the satellite. There is a hypothesis that a tectonic structure of a celestial body is generally controlled by parameters of its orbit(s) (Kochemasov 1999). Unlike planets having one orbit, satellites have two orbits. This increases a complexity or heterogeneity of a tectonic structure of a satellite as compared to a planet (Kochemasov 2006). The lack of topographically manifested helical structures on the Moon may be explained by this proposal (Kochemasov 2008, personal communication).

4 Conclusions

The paper examines capabilities of digital terrain modelling to analyse global topography of Mars, Venus, and the Moon. For the celestial bodies, global digital models and maps of fifteen local, regional, and combined topographic variables were derived. Global maps of topographic variables represented peculiarities of the relief in different ways, according to the physical and mathematical sense of a particular variable. A preliminary analysis demonstrated that the most useful were maps of horizontal and vertical curvatures as well as catchment and dispersive areas. The horizontal curvature revealed flow structures, while the vertical curvature detected mega-scarps. Mapping of the catchment and dispersive areas delineated global networks of ridges and valleys. On binary maps of the catchment area, it was possible to detect several helical structures encircling Mars and Venus from pole to pole. Detailed interpretation of the tectonic helices is the objective of a further research. *TI* map may be used to study the spatial distribution of the water ice in the Martian soil.

Relatively low accuracy of the initial data for Venus and the Moon limited the detection of topographic structures and interpretation of the morphometric maps. Considering current trends in the Solar system investigation, in the medium-term future one would expect the production of new global DEMs of Venus, the Moon, and other celestial bodies (Haruyama et al. 2008), which accuracy would be comparable with the DEM of Mars.

The existed geological maps of Mars, Venus, and the Moon (U.S. Geological Survey 1972; Scott and Carr 1978; Ivanov 2008) can be integrated with the maps of topographic variables. They may be useful in solving various tasks of planetary science: to refine borders of topographic and geological structures, to describe them quantitatively, to analyse their spatial distribution over the planetary surface, etc. The use of digital terrain modelling in planetary studies can tangibly improve our knowledge about planets, satellites, and asteroids.

Acknowledgments

The author thanks GG Dikevich (Moscow, Russia), VF Pankin (Cartgeocentre, Moscow, Russia), RL Kirk (Astrogeology Team, USGS, Flagstaff, USA), AE Fedorov (Annual Seminar ‘The System of the Planet Earth: Non-Conventional Problems of Geology’, Moscow, Russia), GG Kochemasov (Institute of Ore Deposit Geology, Petrography, Mineralogy and

Geochemistry, Russian Academy of Sciences, Moscow, Russia), and three anonymous referees for suggestions, as well as S Slavney (PDS Geosciences Node, Washington University, St. Louis, USA), VG Karpova, EK Vershinina (Lundbeck, Kiev, Ukraine), and TE Petrova (Institute of Mathematical Problems of Biology, Russian Academy of Sciences, Pushchino, Russia) for assistance.

References

Albee AL (2000) Mars 2000. Ann Rev Earth Plan Sci 28:281–304

Baker VR (2001) Water and the Martian landscape. Nature 412:228–236

Bandfield JL (2007) High-resolution subsurface water-ice distributions on Mars. Nature 447:64–67

Barnouin-Jha OS, Cheng AF, Gaskell RW (2008) The surface roughness of asteroid 25143 Itokawa and 433 Eros. In: Abstracts of the 39th Lunar planetary Science Conference, 10–14 March 2008. League City, Texas, # 1297

Beven KJ, Kirkby MJ (1979) A physically-based variable contributing area model of basin hydrology. Hydrol Sci Bull 24:43–69

Boynton WV, Feldman WC, Squyres SW, Prettyman T, Brückner J, Evans LG, Reedy RC, Starr R, Arnold JR, Drake DM, Englert PAJ, Metzger AE, Mitrofanov I, Trombka JI., d'Uston C, Wänke H, Gasnault O, Hamara DK, Janes DM, Marcialis RL, Maurice S, Mikheeva I, Taylor GJ, Tokar R, Shinohara C (2002) Distribution of hydrogen in the near-surface of Mars: evidence for sub-surface ice deposits. Science 297:81–85

Bue BD, Stepinski TF (2006) Automated classification of landforms on Mars. Comp Geosci 32:604–614

Byrne CJ (2007) A large basin on the near side of the Moon. Earth Moon Planets 101:153–188

Cabrol NA, Grin EA (2005) Ancient and recent lakes on Mars. In: Tokano T (ed) Water on Mars and life. Springer, Berlin, pp 235–259

Carr MH, Head III JW (2003) Oceans on Mars: An assessment of the observational evidence and possible fate. J Geophys Res E108:5042

Cochrane CG, Ghail RC (2006) Topographic constraints on impact crater morphology on Venus from high-resolution stereo synthetic aperture radar digital elevation models. J Geophys Res 111:E04007

Fishbaugh KE, Head III JW (2001) Comparison of the north and south polar caps of Mars: new observations from MOLA data and discussion of some outstanding questions. Icarus 154:145–161

Fisher D (1993) If Martian ice caps flow: ablation mechanisms and appearance. Icarus 105:501–511

Florinsky IV (1998a) Combined analysis of digital terrain models and remotely sensed data in landscape investigations. Progr Phys Geogr 22:33–60

Florinsky IV (1998b) Derivation of topographic variables from a digital elevation model given by a spheroidal trapezoidal grid. Int J Geogr Inf Sci 12:829–852

Florinsky IV (2002) Errors of signal processing in digital terrain modelling. Int J Geogr Inf Sci 16:475–501

Florinsky IV (2005) Artificial lineaments in digital terrain modelling: can operators of topographic variables cause them? Math Geol 37:357–372

Florinsky IV (2008) Global lineaments: application of digital terrain modelling. In: Zhou Q, Lees B, Tang G-a (eds) Advances in digital terrain analysis. Springer, Berlin, pp 365–382

Florinsky IV, Grokhlina TI, Mikhailova NL (1995) LANDLORD 2.0: the software for analysis and mapping of geometrical characteristics of relief (in Russian). Geod Cartogr 5:46–51

Ford PG (1992) Magellan global topography, emissivity, reflectivity, and slope data, MGN-V-RDRS-5-GDR-TOPOGRAPHIC-V1.0, MGN-V-RDRS-5-GDR-EMISSIVITY-V1.0, MGN-V-RDRS-5-GDR-REFLECTIVITY-V1.0, and MGN-V-RDRS-5-GDR-SLOPE-V1.0, NASA planetary data system, http://pds-geosciences.wustl.edu/missions/magellan/gxdr/index.htm

Fukue K, Shimoda H, Sakata T (1981) Complete lineament extraction with the aid of shadow-free Landsat image. In: Machine processing of remotely sensed data with special emphasis on range, forest, and wetlands assessment. Proc 7th Int Symp, 23-26 June, 1981. Purdue University, West Lafayette, pp 94–102.

Haruyama J, Ohtake M, Matunaga T, Morota T, Honda C, Torii M, Yokota Y, Ogawa Y, Abe M, Hara S, Hioki K, LISM working group (2008) Kaguya (Selene) / Terrain Camera initial results and perspectives. In: Abstr 37th Lunar planetary sci conf, 13–17 March 2006, League City, Texas, # 1132

Hobbs WH (1904) Lineaments of Atlantic Border region. Geol Soc Amer Bull 15:483–506

Howard AD (1978) Origin of the stepped topography of the Martian poles. Icarus 34:581–599

Howard AD (2000) The role of Eolian processes in forming surface features of the martian polar layered deposits. Icarus 144:267–288

Ivanov MA (2008) Global geological map of Venus: preliminary results. In: Abstr 39th Lunar planetary sci conf, 10–14 March 2008, League City, Texas, # 1017

Jenson SK (1991) Applications of hydrologic information automatically extracted from digital elevation models. Hydrol Proc 5:31–44.

Kirk RL, Howington-Kraus E, Redding B, Galuszka D, Hare TM, Archinal BA, Soderblom LA, Barrett JM (2003) High-resolution topomapping of candidate MER landing sites with Mars Orbiter Camera narrow-angle images. J Geophys Res 108:8088

Kochemasov GG (1999) Theorems of wave planetary tectonics. Geophys Res Abstr 1:700

Kochemasov GG (2006) Anomalous parts on the Lunar crater size – frequency curve, their prediction and comparison of the Lunar and Solar discs granulation patterns. In: Abstr 36th COSPAR Sci Assembly, 16–23 July 2006, Beijing, China. CD-ROM, # 793

Krasilnikov NN, Krasilnikova OI, Shelepin YE (2000) Perception of achromatic, monochromatic, pure chromatic, and chromatic noisy images by real human-observer under threshold conditions. Proc SPIE 3981:78–85.

Kuzmin RO (2005) Ground ice in the Martian regolith. In: Tokano T (ed) Water on Mars and life. Springer, Berlin, pp 155–189

Lattman LH (1958) Technique of mapping geologic fracture traces and lineaments on aerial photographs. Photogram Eng 24:568–576

Liang C, Mackay DS (2000) A general model of watershed extraction and representation using globally optimal flow paths and up-slope contributing areas. Int J Geogr Inf Sci 14:337–358.

Lucey P, Spudis PD, Zuber M, Smith D, Malaret E (1994) Topographic-compositional units on the Moon and the early evolution of the lunar crust. Science 266:1855–1858

Makarov VI (1981) Lineaments: problems and trends of studies by remote sensing techniques (in Russian). Izv Vuz, Geol Razv, 4:109–115

Martz LW, de Jong E (1988) CATCH: a Fortran program for measuring catchment area from digital elevation models. Comp Geosci 14:627–640

Mitrofanov IG (2005) Global distribution of subsurface water measured by Mars Odyssey. In: Tokano T (ed) Water on Mars and life. Springer, Berlin, pp 99–128

Molloy I, Stepinski TF (2007) Automatic mapping of valley networks on Mars. Comp Geosci 33:728–738

Moore ID, Grayson RB, Ladson AR (1991) Digital terrain modelling: a review of hydrological, geomorphological and biological applications. Hydrol Proc 5:3–30

Morozov VP (1979) A course in spheroidal geodesy. 2nd enl. rev. edn (in Russian). Nedra, Moscow

Neukum G, Jaumann R, Hoffmann H, Hauber E, Head JW, Basilevsky AT, Ivanov BA, Werner SC, van Gasselt S, Murray JB, McCord T, the HRSC Co-Investigator Team (2004) Recent and episodic volcanic and glacial activity on Mars revealed by the High Resolution Stereo Camera. Nature 432:971–979

Nimmo F, Pappalardo RT, Giese B (2003) On the origins of band topography, Europa. Icarus 166:21–32

Nozette S, Rustan P, Pleasance LP, Horan DM, Regeon P, Shoemaker EM, Spudis PD, Acton CH, Baker DN, Blamont JE, Buratti BJ, Corson MP, Davies ME, Duxbury TC, Eliason EM, Jakosky BM, Kordas JF, Lewis IT, Lichtenberg CL, Lucey PG, Malaret E, Massie MA, Resnick JH, Rollins CJ, Park HS, McEwen AS, Priest RE, Pieters CM, Reisse RA, Robinson MS, Smith DE, Sorenson TC, Vorder Breugge RW, Zuber MT (1994) The Clementine mission to the Moon: scientific overview. Science 266:1835–1839

Nyrtsov MV (2000) Developing map projections of actual surfaces of celestial bodies and methods of their study. Abstract of Ph.D. thesis (in Russian). Moscow State University of Geodesy and Cartography, Moscow

O'Leary DW, Friedman JD, Pohn HA (1976) Lineament, linear, lineation: some proposed new standards for old terms. Geol Soc Amer Bull 87:1463–1469

Quinn P, Beven K, Lamb R (1995) The ln(a/tan β) index: how to calculate it and how to use it within the TOPMODEL framework. Hydrol Proc 9:161–182

Rappaport NJ, Konopliv AS, Kucinskas AB, Ford PG (1999) An improved 360 degree and order model of Venus topography. Icarus 139:19–31

Sadovnichy VA, Serbenyuk SN, Belov VP, Komissarov VV, Musin OR, Novakovsky BA, Sytenko IN, Usikov DA (1991) Digital model of Phobos surface (in Russian, with English abstract). Vest Mosc Univ, Geogr Ser 2:43–54

Saunders RS, Pettengill GH (1991) Magellan: mission summary. Science 252:247–249

Scott DH, Carr MH (1978) Geologic map of Mars, I-1083. Scale 1:25,000,000. US Geological Survey, Reston

Seidelmann PK, Abalakin VK, Burša M, Davies ME, de Bergh C, Lieske JH, Oberst J, Simon JL, Standish EM, Stooke P, Thomas PC (2002) Report of the IAU/IAG Working Group on cartographic coordinates and rotational elements of the planets and satellites: 2000. Cel Mech Dyn Astr 82:83–111

Shary PA (1995) Land surface in gravity points classification by complete system of curvatures. Math Geol 27:373–390

Shary PA, Sharaya LS, Mitusov AV (2002) Fundamental quantitative methods of land surface analysis. Geoderma 107:1–32

Shults SS (1970) Lineaments (in Russian). Vest Leningr Univ, Geol Geogr Ser 24:50–56

Simons M, Hager BH, Solomon SC (1994) Global variations in the geoid/topography admittance of Venus. Science 264:798–803

Sjogren WL (1997) Magellan spherical harmonic and gravity map data V1.0, MGN-V-RSS-5-GRAVITY-L2-V1.0, NASA planetary data system, http://pds-geosciences.wustl.edu/missions/magellan/shadr_topo_grav/index.htm

Slyuta EN, Kudrin LV, Sinilo VP (1989) Preliminary data on the nature of a planetary system of lineaments observed in radar images of Venus (data from Venera-15 and -16). Cosmic Res 27:786–797

Smith DE (2003) MGS MOLA mission experiment gridded data record, MGS-M-MOLA-5-MEGDR-L3-V1.0, NASA planetary data system, http://pds-geosciences.wustl.edu/missions/mgs/megdr.html

Smith DE, Zuber MT, Neumann GA, Lemoine FG (1997) Topography of the Moon from the Clementine lidar. J Geophys Res E102:1591–1611

Smith DE, Sjogren WL, Tyler GL, Balmino G, Lemoine FG (1999) The gravity field of Mars: results from Mars Global Surveyor. Science 286:94–97

Smith DE, Zuber MT, Solomon SC, Phillips RJ, Head JW, Garvin JB, Banerdt WB, Muhleman DO, Pettengill GH, Neumann GA, Lemoine FG, Abshire JB, Aharonson O, Brown CD, Hauck SA, Ivanov AB, McGovern PJ, Zwally HJ, Duxbury TC (1999) The global topography of Mars and implications for surface evolution. Science 284:1495–1503

Solomon SC, Head JW, Kaula WM, McKenzie D, Parsons B, Phillips RJ, Schubert G, Talwani M (1991) Venus tectonics: initial analysis from Magellan. Science 252:297–312

Speight JG (1974) A parametric approach to landform regions. In: Brown EH, Waters RS (eds) Progress in geomorphology: papers in honour of DL Linton. Institute of British Geographers, London, pp 213–230

Stepinski TF, Stepinski AP (2005) Morphology of drainage basins as an indicator of climate on early Mars. J Geophys Res 110:E12S12

Swets JA (1961) Detection theory and psychophysics: a review. Psychometrika 26:49–63

Takahashi H (1981) A lineament enhancement technique for active fault analysis In: Machine processing of remotely sensed data with special emphasis on range, forest, and wetlands assessment. Proc 7th Int Symp, June 23-26, 1981. Purdue University, West Lafayette, pp 103–112

Tokano T (2005) Water cycle in the atmosphere and shallow subsurface. In: Tokano T (ed) Water on Mars and life. Springer, Berlin, pp 191–216

Trifonov VG, Makarov VI, Safonov YG, Florensky PV (eds) (1983) Space remote-sensing data in geology (in Russian, with English contents). Nauka, Moscow

US Geological Survey (1972) Geologic atlas of the Moon. Scale 1:1,000,000. US Geological Survey, Washington

Vincenty T (1975) Direct and inverse solutions of geodesics on the ellipsoid with application of nested equations. Surv Rev 23:88–93

Watters TR, Cook AC, Robinson MS (2001) Large-scale lobate scarps in the southern hemisphere of Mercury. Plan Space Sci 49:1523–1530

Weijermars R (1985/86) The polar spirals of Mars may be due to glacier surges deflected by Coriolis forces. Earth Plan Sci Let 76:227–240

Wieczorek MA (2007) Gravity and topography of the terrestrial planets. In: Spohn T (ed) Treatise on geophysics, vol 10. Elsevier, Amsterdam, pp 165–206

Williams DA, Greeley R, Hauber E, Gwinner K, Neukum G (2005) Erosion by flowing Martian lava: New insights for Hecates Tholus from Mars Express and MER data. J Geophys Res 110:E05006

Wilson JP, Gallant JC (2000) Digital terrain analysis. In: Wilson JP, Gallant JC (eds) Terrain analysis: principles and applications. Wiley, New York, pp 1–28

Zlatopolsky AA (1992) Program LESSA (lineament extraction and stripe statistical analysis): automated linear image features analysis – experimental results. Comp Geosci 18:1121–1126.

Zuber MT (1996) Clementine lunar topography V1.0, CLEM1-L-LIDAR-5-TOPO-V1.0. NASA planetary data system, http://pds-geosciences.wustl.edu/missions/clementine/gravtopo.html

Zuber M, Smith DE, Lemoine FG, Neumann G (1994) The shape and internal structure of the Moon from the Clementine mission. Science 266:1839–1843

Zuber MT, Solomon SC, Phillips RJ, Smith DE, Tyler GL, Aharonson O, Balmino G, Banerdt WB, Head JW, Johnson CL, Lemoine FG, McGovern PJ, Neumann GA, Rowlands DD, Zhong S (2000) Internal structure and early thermal evolution of Mars from Mars Global Surveyor topography and gravity. Science 287:1788–1793

CHAPTER 9

Scalability of Techniques for Online Geographic Visualization of Web Site Hits

Nigel Stanger

Department of Information Science, University of Otago, Dunedin, New Zealand

Abstract

Extremely large data sets are now commonplace, and they are often visualized through the World Wide Web. Scalability of web-based visualization techniques is thus a key issue. This paper investigates the scalability of four representative techniques for dynamic map generation and display (e.g., for visualizing geographic sources of web site hits): generating a single composite map image, overlaying images on an underlying base map and two variants of overlaying HTML on a base map. These four techniques embody a mixture of different display technologies and distribution styles (three server-side and one distributed across both client and server). Each technique was applied to 20 synthetic data sets of increasing size, and the data set volume, elapsed time and memory consumption were measured. The results show that all four techniques are suitable for small data sets comprising a few thousand points, but that the two HTML techniques scale to larger data sets very poorly across all three variables.

Keywords: web mapping, scalability, dynamic map generation, visualization, geolocation, distribution style, World Wide Web, Google Maps

1 Introduction

When administering a Web site, it is normal to analyze the nature of traffic to the site. Information on the geographic sources of traffic can be particularly useful in the right context. For example, e-commerce sites might wish to determine the geographical distribution of visitors, so as to decide where best to target marketing resources. One solution is to plot visitors' geographical locations on a map. Geographical information systems (GIS) were already being used for these purposes prior to the advent of the World Wide Web (Beaumont 1991), and it is natural to extend these ideas to online visualization of Web site traffic. Indeed, this is a classic example of applying geographic visualization techniques to data sourced from the "virtual world" (Kitchin and Dodge 2002, p 344).

However, scalability is an issue for such visualizations, with rapid growth in both Internet traffic and data collected leading to potentially very large data sets (Andrienko et al. 2005, p 107), which are commonly accessed via the Web. Multi-terabyte data warehouses are now commonplace (Babcock 2006), and the data set underlying Google Earth (a Web-based application) comprises at least 70 terabytes of raw images and 500 gigabytes of index data (Chang et al. 2006; Bar-Zeev 2007). Even a moderately busy Web site can easily generate millions of hits—and thus megabytes of data—per day. This implies a strong need for scalable techniques to visualize such data on the Web; an issue that no Web cartographer should ignore.

Despite the clear importance of scalability in this context, there appears to have been little work on testing the performance bounds of the many Web-based cartographic visualization techniques. Ideally, a technique should not only efficiently fulfil the task of plotting data points on a map, but also provide tangible benefits to end-users. Scalability is a key issue for Web applications in general (Offutt 2002, p 28), and for online activity visualization in particular (Eick 2001, p 50), so techniques that can scale to a large number of points are of particular interest.

In this paper we investigate the scalability of four representative techniques (server-side image generation, server-side image overlay, server-side HTML overlay and distributed HTML overlay, as exemplified by Google Maps) for online Web visualization of the geographic sources of downloads from a web site. General background information about the research and the general classes of technique available is provided in Sect. 2, and the selection process and details of the techniques chosen for testing are discussed in Sect. 3.

The scalability of the four techniques was tested to determine how well each technique handled large numbers of data points. A series of experiments was conducted on each technique with progressively larger synthetic data sets, and the data set volume, elapsed time and memory consumption were measured. The experimental design is discussed in Sect. 4.

Informal testing suggested that the server-side image generation and server-side image overlay techniques would scale best. This was borne out by the results of the experiments, which show that both techniques scale well to very large numbers of points. The other two techniques proved reasonable for less than about 500–1,000 points, but their performance deteriorated rapidly beyond this. The results are discussed in Sect. 5.

The intent of the experiments was not to identify statistically significant differences in performance across the four techniques. It was expected that variations across techniques would be reasonably clear-cut, and the experiments were designed to test this expectation. However, the two best performing techniques, server-side image generation and server-side image overlay, produced very similar results, so a more formal statistical analysis of these techniques may be warranted. This and other possible future directions are discussed in Sect. 6.

2 Background

This current research arose from implementing a digital institutional repository for the University of Otago School of Business[1] in November 2005 (Stanger and McGregor 2006, 2007), using the GNU EPrints repository management software[2]. This repository quickly attracted interest from around the world and the number of abstract views and full text downloads steadily increased. There was great interest within the University in tracking this traffic, particularly with respect to the geographic sources of hits. The EPrints statistics software developed at the University of Tasmania (Sale and McGee 2006) proved useful in this regard, providing detailed per-eprint and per-country download statistics, as illustrated in Fig. 9.1. However, while the display in Fig. 9.1 provides an ordered ranking of the total hits from each country, it does not include detail below the country level, despite having the data to do so. More importantly, it does not visualize the spatial distribution of hit sources around the globe. The author therefore explored techniques for plotting hit data onto a world map, so as to provide a more cartographic view of the data.

[1] http://eprints.otago.ac.nz/

[2] http://www.eprints.org/

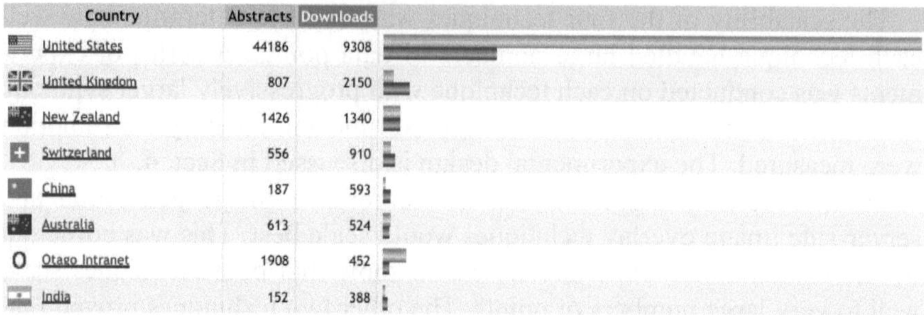

Country	Abstracts	Downloads
United States	44186	9308
United Kingdom	807	2150
New Zealand	1426	1340
Switzerland	556	910
China	187	593
Australia	613	524
Otago Intranet	1908	452
India	152	388

Fig. 9.1. An example of the by-country display generated by the Tasmania statistics software

The Internet has enabled unprecedented access to large data sets. Public access to all kinds of data is now the norm, but effective and equitable access requires lightweight, accessible tools without excessive computing resource or user training needs. This has been an issue with, for example, the development of Web-based public participation GIS (Kingston 2002, pp 108–109) and Web mapping applications (Zhao and Shneiderman 2005, pp 414–415). In this research, therefore, preference was given to techniques that could be used in a modern Web browser without additional client software, so as to cater to the widest possible audience and reduce the impact of wide variation in client hardware and software environments (Offutt 2002, pp 27–28).

There have been several prior efforts to geographically visualize Web activity. Lamm et al. (1996) developed a sophisticated system for real-time visualization of Web traffic on a 3D globe, but this required a virtual reality environment, thus limiting its general applicability. Papadakakis et al. (1998) described a similar system called *Palantir*, which ran as a Java applet within a Web browser. Dodge and Kitchin (2001) describe several other similar systems for mapping Web and Internet traffic.

These early systems suffered from the limitation that there was no public infrastructure for geolocating IP addresses (that is, translating them into geographic coordinates). They generally used `whois` lookups or parsed the domain name in a crude attempt to guess the country of origin (Lamm et al. 1996). Locations outside the United States were typically aggregated by country and mapped to the capital city (Lamm et al. 1996; Papadakakis et al. 1998; Jiang and Ormeling 2000). Reasonably accurate and detailed databases were commercially available (Lamm et al. 1996, p 1466), but a lack of public access limited their utility.

Things have improved in recent years with the advent of freely available and reasonably accurate geolocation services (e.g., Maxmind[3] and IP2Location[4]) with worldwide coverage and city-level resolution. For example, Maxmind's free *GeoLite City* database claims "69% accuracy on a city level for the US within a 25 mile radius" (Maxmind 2008). Their commercial *GeoIP City* database claims 81% accuracy for the same parameters.

The techniques used by these prior systems can generally be divided into two classes. Those of the first class generate a single bitmap image by programmatically plotting points onto a base map image; the composite map image is then displayed at the client. Such techniques shall henceforth be referred to as *single-layer* techniques. Techniques of the second class return a base map image plus one or more overlays of plotted points. The overlay(s) and the base map are composited at the client. This is analogous to the multiple layer construct long used in GIS to arrange data of like geometric type and theme (Longley et al. 2005). This class of techniques shall henceforth be referred to as *multi-layer* techniques.

Both classes are used in the aforementioned systems, but multi-layer techniques appear more prevalent. For example, Palantir used a multi-layer technique, in which a client-side Java applet overlaid graphic elements onto a base map image retrieved from the now-defunct Xerox online map server (Papadakakis et al. 1998). A more recent example is the Google Maps API (Google 2008), which enables Web developers to easily embed interactive maps within Web pages using client-side JavaScript. Google Maps uses a dynamic multi-layer technique that has become feasible with the recent widespread browser support for CSS (Cascading Style Sheets) positioning and Ajax (Asynchronous JavaScript and XML) technologies (Garrett 2005).

Multi-layer techniques enjoy a particular advantage over single-layer techniques, as they can provide a more flexible GIS-like interaction with the map, with multiple layers that can be activated and deactivated as desired. This could explain why such techniques are more common in the literature. However, many Web-based multi-layer techniques rely either on the installation of additional client-side software, or on more recent Web technologies such as CSS and Ajax (Zhao and Shneiderman 2005, p 414). Single-layer techniques typically do not rely on these things, so they should be portable to a wider range of environments.

Each technique comprises a specific technology or collection of technologies (such as transparent bitmap overlays + CSS positioning), implemented

[3] http://www.maxmind.com/

[4] http://www.ip2location.com/

using a specific distribution style. For example, one single-layer technique might be implemented completely server-side while another might use a mixture of server-side and client-side processing. Multi-layer techniques may also adopt different distribution styles, and the overlays might take the form of transparent images, absolutely positioned HTML elements, dynamically generated graphics, etc.

The wide variety of available techniques raises the question of which are most suitable for visualizing large data sets. For example, at the time of writing the Otago EPrints repository had been accessed from over 123,000 distinct IP addresses, each potentially representing a distinct geographical location. Informal testing suggested that a single-layer composite map image performed well with this volume of data, taking at most a few seconds to load and display, whereas Google Maps took several minutes to load and display a map containing only a few thousand points.

For the experiments described here, the range of techniques was first narrowed down to just four: server-side image generation, server-side image overlay, server-side HTML overlay and distributed HTML overlay (using Google Maps). The selection process and details of the techniques chosen are discussed next.

3 Technique Selection

In this section the four techniques chosen for testing are discussed in more detail, along with the reasons for choosing them. First, the impact of distribution style on the choice of technique is discussed. This is followed by a brief examination of how each technique works in practice, its implementation requirements, its relative advantages and disadvantages, and any other issues peculiar to the technique.

3.1 Distribution Style

Wood et al. (1996) and MacEachren (1998) identified four distribution styles for Web-based geographic visualization software. The *data server* style is where the server only supplies raw data, and manipulation, display and analysis all take place at the client, so this is primarily a client-side processing model, as illustrated in Fig. 9.2a. Palantir implemented a multi-layer technique using this distribution style (Papadakakis et al. 1998), by means of a Java applet running at the client. The data server distribution style can provide a dynamic interactive user experience, but clearly requires support for executing application code within the Web browser, using technologies like JavaScript, Java applets or Flash. JavaScript is now integrated into most browsers, but neither Java nor Flash can be guaranteed in a typical browser installation. Java- or Flash-based *data server* techniques are therefore not considered, but JavaScript-based *data server* techniques (such as that used by Google Maps) are feasible.

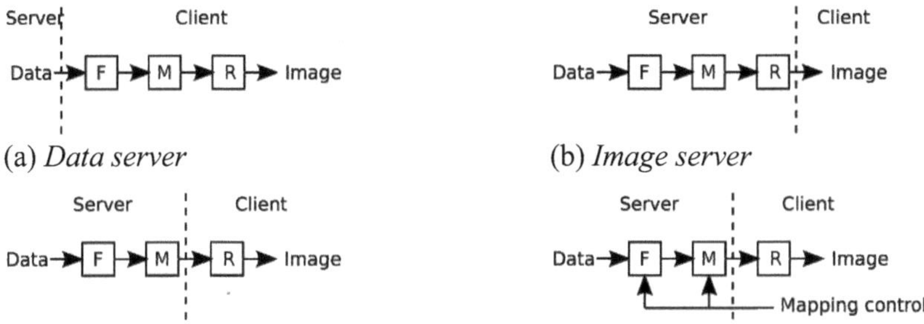

(a) *Data server* (b) *Image server* (c) *Model interaction environment* (d) *Shared environment*

Fig. 9.2. Distribution styles for Web-based geographic visualization (Wood et al. 1996); F = filtering, M = mapping, R = rendering

Conversely, the *image server* style is where the display is created and manipulated entirely at the server and the client is only a passive viewer, so this is primarily a server-side processing model, as illustrated in Fig. 9.2b. Techniques that use this style require no additional client-side software, but the resultant visualization can be static and non-interactive (Cammack 1999, p 159), as it is typically a simple bitmap image.

The *model interaction environment* style is where a model created at the server can be explored at the client, as illustrated in Fig. 9.2c. MacEachren (1998) calls this the "3D model interaction" style, but this does not really fit the current context. Wood et al. (1996) originally applied this distribution style to VRML models for GIS applications, but it could be applied to any interactive model that is generated at the server, then downloaded to and manipulated at the client. "Model interaction environment" therefore seems a more appropriate name. The distinguishing feature of this style is that there is no ongoing interaction between client and server after the model has been downloaded. The downloaded model can be dynamic and interactive, but changing the underlying data requires another round-trip to the server to generate a new model. Similar restrictions apply to techniques using this style as to the data server style, so Java- and Flash-based model interaction environment techniques are not considered, along with solutions like VRML or SVG that require browser plug-ins (although native SVG support is appearing in some browsers).

Finally, the *shared environment* style is where data manipulation occurs at the server, but control of that manipulation, and rendering and display all occur at the client, as illustrated in Fig. 9.2d. This is essentially the *model interaction environment* style plus a feedback loop from client to server, thus enabling a more flexible, dynamic interaction. Ajax technologies can support this kind of distribution style, for example, Sayar et al. (2006) used Ajax to integrate Google Maps into existing GIS visualization Web services. Specific shared environment techniques can be eliminated from consideration based on the same criteria applied previously (e.g., no Java- or Flash-based techniques).

3.2 Single-layer Techniques

Single-layer techniques directly plot geolocated IP addresses onto a base map image, and then display the composite image at the client. A typical example of the output that might be produced is shown in Fig. 9.3. Such techniques require software to programmatically create and manipulate bitmap images, and to transform latitude/longitude coordinates into projected map coordinates.

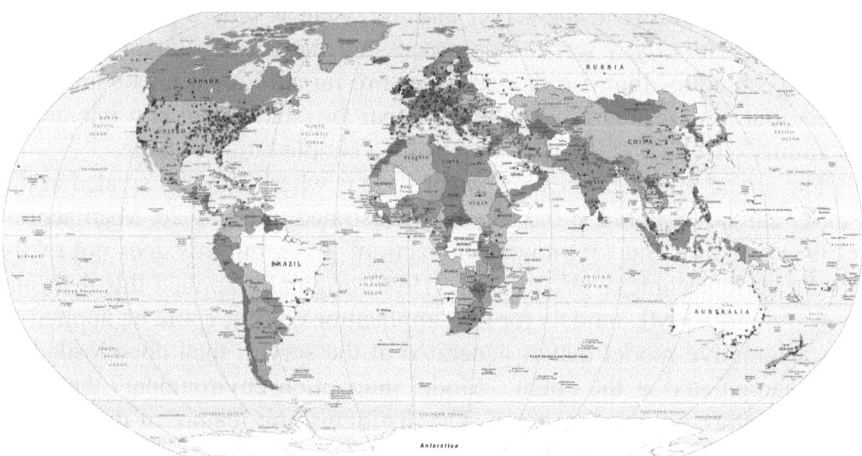

Fig. 9.3. Sample output from the (single-layer) server-side image generation technique; the base map is from the CIA World Factbook (2008)

Single-layer techniques could use any of the distribution styles discussed in Sect. 3.1. However, all but the *image server* style would require additional client-side software for generating images and performing projection transformations, so only single-layer techniques using the *image server* distribution style (or *server-side image generation* techniques) are considered here. For this research, the author chose a representative server-side image generation technique using the GD image library[5] and the PROJ.4 cartographic projections library[6].

Server-side image generation techniques provide some distinct advantages. They are simple to implement and quickly produce the final image. They are also bandwidth efficient (Cammack 1999, p 159), as the size of the final image is determined only by its pixel count and compression method, not the number of points plotted. The volume of data sent to the client should therefore remain relatively constant across data sets.

Server-side image generation techniques have some disadvantages, however. First, a suitable base map image must be acquired. This could be generated from a GIS or licensed from a third party (raising possible copyright issues). Second, the compression method used for the composite map can impact visual quality. Lossy compression methods such as JPEG can make points plotted on the map appear fuzzy or "muddy", even at high quality levels. Lossless compression methods such as PNG avoid this, but can produce larger files. Finally, it is harder to provide interactive map manipulation with server-side image generation techniques (Cammack 1999, p 159), as the output is a static image. Anything that changes the map content (such as panning or hiding points) will force regeneration of the entire image. Limited zooming is possible with a very high-resolution base map image.

3.3 Multi-layer Techniques

Multi-layer techniques indirectly plot points onto the base map by means of one or more independent overlays. This provides a significant advantage over single-layer techniques by providing multiple layers that can be individually manipulated. This is like the multi-layer functionality of a GIS, providing effective interactive visualizations of geographic data (Wood et al. 1996; MacEachren 1998).

Until relatively recently, implementing multi-layer techniques would likely have required additional client-side software, but most modern browsers now support absolute positioning of HTML elements using CSS. This enables the creation of overlays using nothing more than HTML, CSS and a few bitmap images. The author has identified two main multi-layer techniques, which can be termed *image overlay* and *HTML overlay*.

[5] http://www.boutell.com/gd/
[6] http://www.remotesensing.org/proj/

An image overlay comprises a transparent bitmap image into which points are plotted, which is then overlaid on the base map image (Golub and Shneiderman 2003), producing results essentially identical to Fig. 9.3. The overlay image must be in either PNG or GIF format, as JPEG does not support transparency, but the overlay image compresses extremely well because it is mostly "white space". This also eliminates the image quality issue noted earlier. The size of the image overlay should be roughly, but not directly, proportional to the number of points plotted.

As noted in Sect. 3.2, generating images at the client requires additional software, so only the *data server* distribution style will be considered here for image overlays (i.e., *server-side image overlay*). That is, both the base map image and the overlay(s) are generated at the server.

An HTML overlay comprises a collection of HTML elements corresponding to the points plotted, positioned over the base map image using CSS. One option is to use <IMG> elements to place icons on the base map, which is the approach adopted by Google Maps (see Fig. 9.4). Another option is to use appropriately sized, colored and positioned <DIV> elements, again producing results essentially identical to Fig. 9.3.

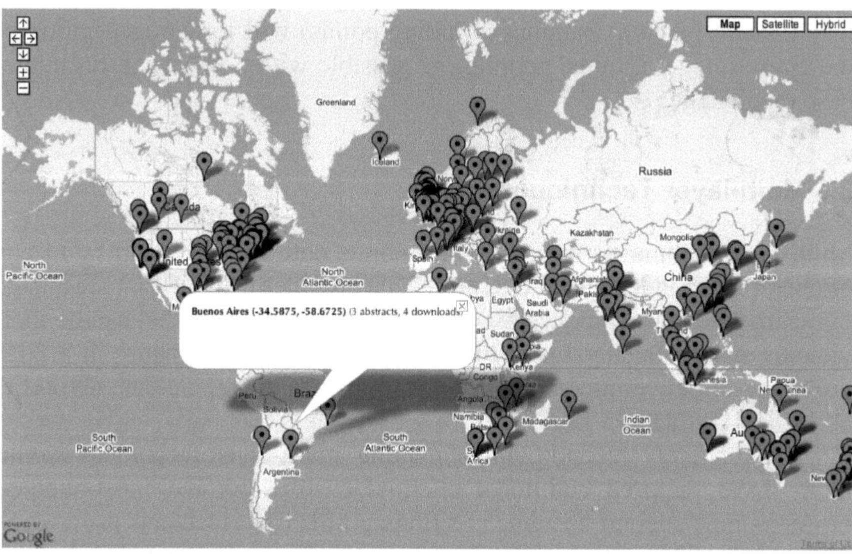

Fig. 9.4. Sample output from the (multi-layer) Google Maps technique

HTML overlays may be generated at either the server or client, and do not require additional client-side software, because only HTML (i.e., text) is being generated. This can be achieved using client-side JavaScript, so HTML overlays can use any of the distribution styles discussed in Sect.

3.1. Two representative HTML overlay techniques were adopted for the experiments: *server-side HTML overlays* (*image server* distribution style) and *distributed HTML overlays* (*data server* distribution style). The latter is exemplified by Google Maps, so from now on we shall refer to this technique as "Google Maps", to distinguish it more clearly from server-side HTML overlays. Since Google Maps uses `<IMG>` elements, `<DIV>` elements were used for the server-side HTML overlay.

Server-side HTML overlays are simple to implement: all that is required is code to perform projection transformations and generate corresponding `<DIV>` elements. Google Maps (Google 2008) is a more complex proposition. It uses the *data server* distribution style, running JavaScript within the client browser to manipulate the base map and its overlays. Data and map images are requested asynchronously from the server as needed. (This latter point may seem to imply that Google Maps uses the *shared environment* rather than the *data server* distribution style. However, with Google Maps the server is merely a passive supplier of data and does not generate or manipulate the map.)

The primary advantage of Google Maps is the rich functionality it provides for map generation and interaction. Users may pan and zoom within broad limits. Satellite imagery and terrain layers are also available. Details about each point can be displayed in callouts, as shown in Fig. 9.4. Google Maps also has a proven record for visualizing and managing network resources, such as worldwide computing grids (Gibbins and Buyya 2006).

However, Google Maps also has some significant disadvantages[7]. First, it is a distributed application, making it complex to implement, test and debug (Enslow 1978; Bates 1995). Second, the web server must have a registered API key, which is verified with Google every time a page attempts to use the API. The client must also connect to Google's servers to download the JavaScript code, meaning that an active Internet connection is essential even when client and server are running on the same machine.

The most significant disadvantage of all HTML overlay techniques (including Google Maps), however, is that the size of the overlay is directly proportional to the number of points, so a very large data set will generate an even larger amount of HTML (Cammack 1999, pp 158–159). Large data sets may therefore cause excessive browser memory consumption, implying that these techniques will not scale well at the high end. However, they may still be appropriate for small data sets that require interactive manipulation. Encoding methods such as that suggested by Zhao and Shneiderman (2005) may also help.

[7] Interestingly, the Google Earth application addresses many of these issues, but since it is not a browser-based solution it is not considered here.

4 Experimental Design

A series of experiments was undertaken to test the scalability of the four chosen techniques, using a collection of progressively larger synthetic data sets. The first data set comprised one point at the South Pole. A regular grid of points at one-degree intervals was constructed by incrementing latitude and longitude, with each data set twice the size of its predecessor. A total of twenty-one data sets were created, ranging from one to 1,048,576 ($=2^{20}$) points. A plot of the 16,384-point data set is shown in Fig. 9.5.

We must immediately acknowledge that the map in Fig. 9.5 will probably seem quite bizarre to most cartographers. Certainly this is not a realistic use case, and a real application would at the very least use a density reduction method such as a common dot map (Dent 1990). However, the aim of these experiments was explore the boundaries of performance for Web mapping applications, and it makes no difference computationally how data points are geographically distributed. For a given technique, 16,384 points will take the same time to plot regardless of their location, so the results of the experiments can be generalized to any data set. We chose synthetic data sets as they were easier to generate and their parameters could be tightly controlled.

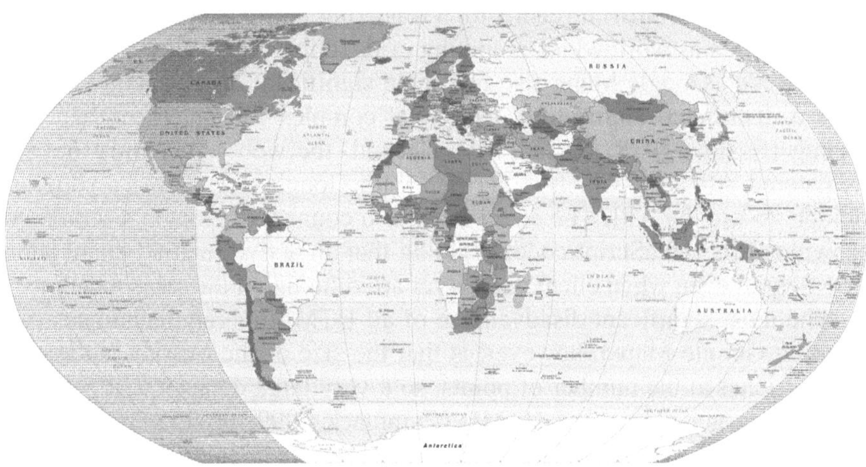

Fig. 9.5. The 16,384-point synthetic data set plotted on the base map

The grid spacing used meant that 64,800 points "filled" the entire map, so the five largest data sets had many overlapping points. This is normal for real download data, as there are often multiple hits from the same location, but note again that in a real application we would aggregate these duplicates in some way.

The focus on scalability implied key variables of page load time (subdivided into time taken to generate map data, time taken to transfer map data and related material to the client across the network, and time taken by the client to display the map), browser memory consumption and volume of data generated (which impacts on both storage and network bandwidth).

Unfortunately, as noted in Sect. 3.3, Google Maps requires an active Internet connection, so the experiments could not use an isolated network. Local network traffic was thus a potential confounding factor, which was eliminated by running both server and client on the same machine[8]. This also enabled independent measurement of data generation and page display times, thus simplifying the data collection process and ensuring that the client and server processes did not unduly interfere with each other.

Internet performance could arguably still have a confounding effect on Google Maps, but this would likely only affect the initial API download (about 235 kB), which would be locally cached thereafter. API key verification occurs on every page load, but the data volume is very small, so it is less likely to be affected by Internet issues. Any such issues would be immediately apparent, as it would simply block the server from proceeding.

For each data set, its size, the time taken to generate it, the time taken to display the resultant map in the browser, and the amount of real and virtual memory consumed by the browser were recorded. It was also intended to measure server memory consumption, but this proved difficult to isolate, and was dropped. Each test was run up to twenty times, where feasible, to compensate for random variations; some tests were run fewer times because they took an excessive amount of time. Testing for a technique was generally halted when a single test run took longer than about five minutes, as by then performance had already deteriorated well beyond usable levels. The Web browser was shut down and reloaded before each group of tests.

4.1 Technique Implementation

As noted in Sects. 3.2 and 3.3, the server-side image generation, server-side image overlay and server-side HTML overlay techniques were all implemented using the *image server* distribution style, while the Google Maps technique was implemented using the *data server* distribution style.

Fig. 9.6 shows the overall architecture for the implementation of each technique. Browser requests were handled by a server-side combination of PHP and Perl scripts that generated an appropriate response for each

[8] A Power Macintosh G5 1.8 GHz (single processor) with 1 GB RAM, running Mac OS X 10.4.7, Apache 2.0.55, PHP 4.4 and Perl 5.8.6.

technique. The first three techniques shared the same base map (which used the Robinson projection), taken from a map collection released into the public domain by the CIA (2008). All three techniques used the PROJ.4 cartographic projections library, and the first two techniques used the GD graphics library.

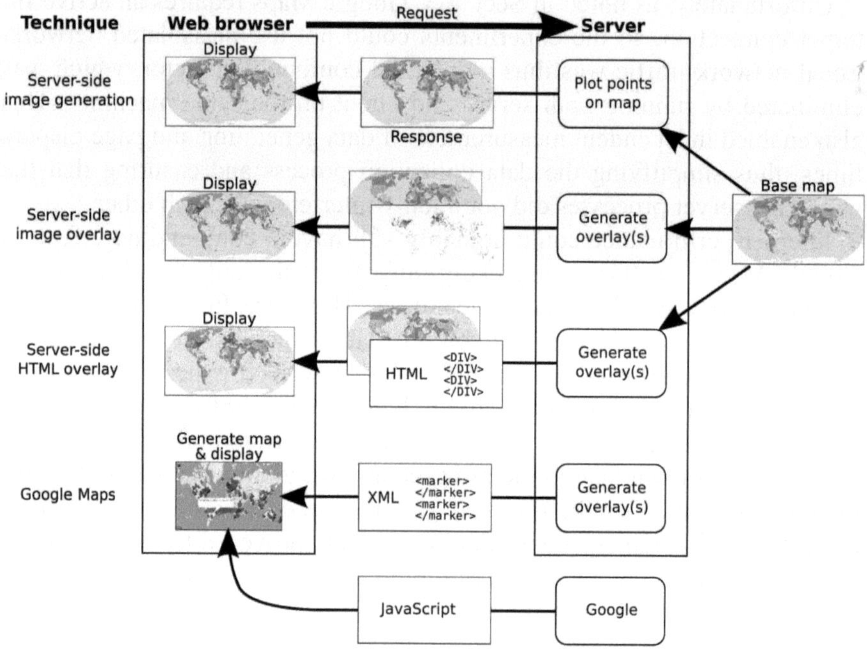

Fig. 9.6. Technique implementation architecture

The Google Maps technique differed from the others in that the server generated no images. Instead it generated and returned an XML data set containing the data points, which were then plotted by the browser using the Google Maps JavaScript API. Google Maps currently only supports the Mercator projection, but the API automatically handles projection transformations.

One obvious concern is that three of the techniques use the Robinson projection—simply because of the ready availability of a high-quality base map image using that projection (CIA 2008)—while the fourth uses the Mercator projection. The consequent differences in projection transformations could arguably invalidate cross-technique comparisons. Informal testing with the PROJ.4 library found only a slight difference in performance—less than 1% over 3 million transformations—between the Mercator and Robinson projection transformations. This implies that the transformations for both projections are computationally similar. Similar testing with

Google Maps, however, showed it was about 18 times slower than PROJ.4. Inspection of the JavaScript source revealed nothing unusual about the implementation of the Mercator transformation, so the difference is most likely attributable to the use of compiled C code versus interpreted JavaScript. The performance figures could be normalized to make them directly comparable, but this would only make sense in the context of a full statistical analysis. The goal of the experiments is to explore the boundaries of real-world performance, so the fact that Google Maps is 18 times slower is clearly important and should be taken into account.

5 Results and Discussion

As noted above, the goal of these experiments was not a full statistical analysis of the performance of the different techniques, but rather identifying broad trends. The remainder of this section discusses in detail the results for data volume, page load time and memory consumption. All plots use log-log scales.

5.1 Data Volume

The data generated by the server for each data set was saved to a file and its size in bytes recorded. For the server-side image generation and server-side image overlay techniques, the file was a bitmap image; whereas for the server-side HTML overlay and Google Maps techniques, the file comprised HTML or XML text, respectively.

There was also a certain amount of fixed overhead associated with each technique, summarized in Table 9.1. This comprised static files that were always downloaded to the client, such as the base map image, various icons, the base source for the Web page and the JavaScript source for the Google Maps API.

The volume of data generated by each technique, including fixed overhead, is shown in Fig. 9.7 (note that "server-side" has been omitted from the technique names in this and subsequent figures). There is an immediately apparent divergence between the techniques that generate images (server-side image generation and server-side image overlay) and the techniques that generate text (server-side HTML overlay and Google Maps).

Both the server-side image generation and server-side image overlay techniques scale well with regard to data volume. Interestingly, the data volume of the image generation technique increases by about 8 kB up to 8,192 points, but then *drops* by about 90 kB over the next three data sets.

This is because at this point the number of points plotted covers much of the base map. A large portion of the composite map image is thus a single color (see Fig. 9.5), which compresses more efficiently.

Table 9.1. Fixed overhead for each technique; the largest contributing item for each technique is shown in **bold face**

Technique	Fixed overhead	Content
Server-side image generation	629 bytes	**Web page source**
Server-side image overlay	≈181 kB	Web page source **base map image (JPEG)**
Server-side HTML overlay	≈181 kB	Web page source **base map image (JPEG)**
Google Maps	≈235 kB	Web page source base map image tiles (PNG) **API (JavaScript)** various icons (PNG)

The data volume of the image overlay technique appears constant, but actually increases by about 2 kB across the range. This has important implications for handling multiple layers. Because the overlay images are small (less than 2 kB for one million points), it should be feasible to preload several overlay images on the client and interactively toggle them.

The server-side HTML overlay and Google Maps techniques clearly do not scale well with respect to data volume, and visibly diverge from the other two techniques once the data volume exceeds about 5% of the fixed overhead. For HTML overlay this occurs somewhere between 64 and 128 points, and for Google Maps somewhere between 256 and 512 points. The divergence increases rapidly for both techniques, with HTML overlay suffering most. The latter is because HTML overlay generates additional CSS attributes (i.e., more text) to position the `<DIV>` elements, whereas Google Maps returns a more compact list of latitude/longitude coordinates.

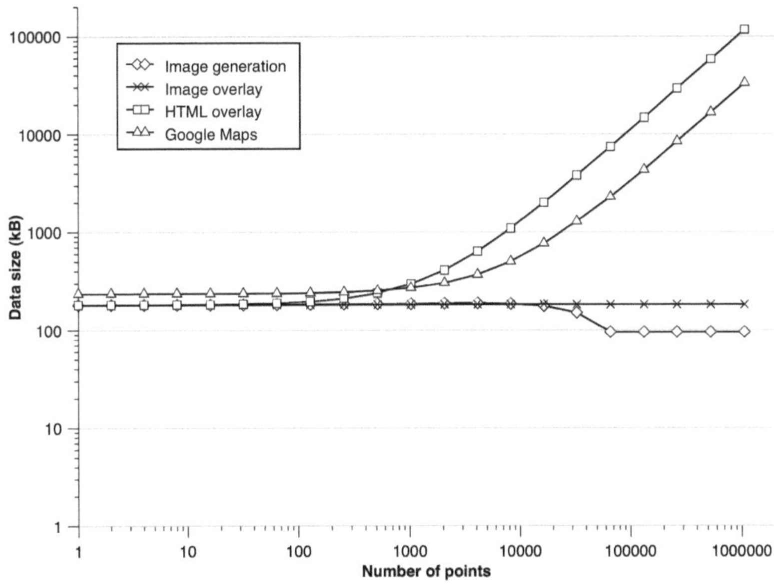

Fig. 9.7. Comparison of generated data size for each technique (log-log scale)

5.2 Page Load Time

For each test run, both the time taken to generate the data at the server and to display the page in the client browser were recorded. The former is shown in Fig. 9.8, the latter in Fig. 9.9, and the combined time in Fig. 9.10.

5.2.1 Data Generation Time

The results (see Fig. 9.8) show that the time taken to generate the source data increases in proportion to the number of points to be plotted. It is interesting to note that data generation for the two techniques that generate text is generally faster than for two techniques that generate images.

Server-side image generation generally takes the longest to generate its data. This is because it has to perform projection transformations, plot points onto the base map image and compress the composite image as a JPEG. The image to be compressed is also moderately complex, adding to the data generation time. Server-side image overlay performs somewhat better because it uses a less expensive compression method (PNG) and the image to be compressed is much simpler (a collection of colored dots on a blank background).

Server-side HTML overlay appears faster at generating data than either image-generating technique at the low end, but performs similarly at the high end. In this technique the server only performs projection transformations; no images are generated and there is no compression. At the high end, however, this advantage is offset by the significant data volumes generated. Google Maps is faster again, because almost all processing occurs at the client; the server's only role is to generate a list of coordinates.

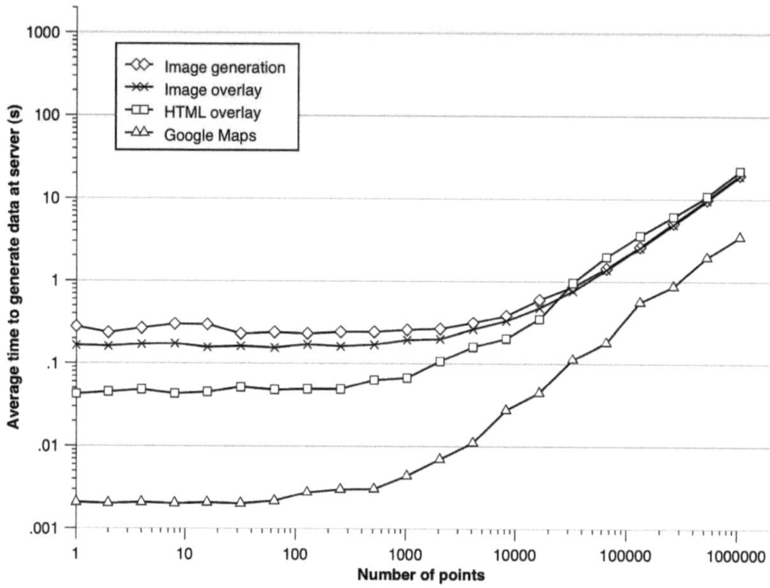

Fig. 9.8. Comparison of data generation time for each technique (log-log scale)

In terms of data generation, all four techniques appear to scale reasonably well. The image-generating techniques perform worse at the low end because they require more complex processing, but this is offset at the high end by the relatively constant data volume. Conversely, the text-generating techniques perform better at the low end, but are negatively impacted at the high end by the sheer volume of data (tens or hundreds of megabytes vs. hundreds of kilobytes).

5.2.2 Map Display Time

The results (see Fig. 9.9) reveal spectacular differences between the image-generating and text-generating techniques. The map display time is essentially constant for both image-generating techniques, which is not surprising given that the data volume is also essentially constant, and that the browser is simply loading and displaying static images. Image overlay appears slightly slower than image generation, probably because image overlay loads two images from the server instead of image generation's one image.

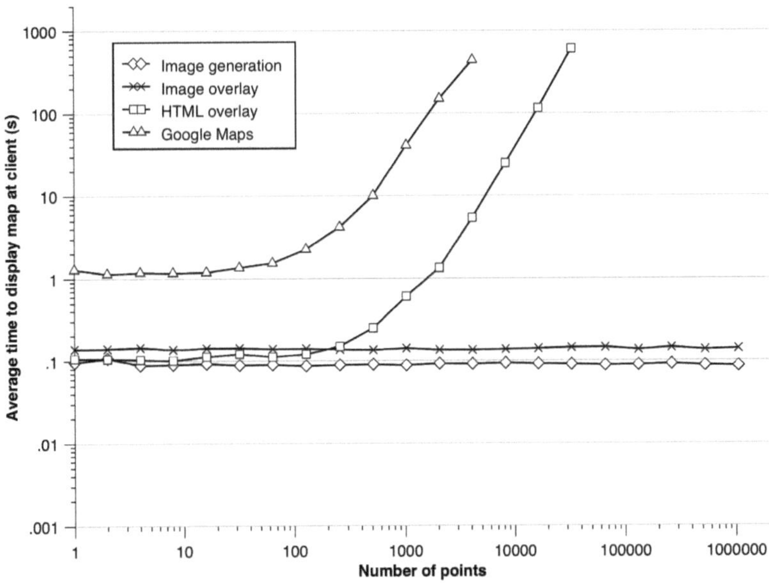

Fig. 9.9. Comparison of map display time for each technique (log-log scale)

In contrast, the text-generating techniques clearly do not scale well with respect to map display time. Google Maps suffers most, with display time exceeding ten seconds shortly past 512 points. Testing was abandoned at 4,096 points, which took over seven minutes. HTML overlay fares better, exceeding ten seconds somewhere between 4,096 and 8,192 points. Testing was abandoned at 32,768 points, which took almost ten minutes.

5.2.3 Combined Time

Combining the data generation and map display times (see Fig. 9.10) yields little change to the curves for the text-generating techniques, because the data generation times are very small compared to the map display times. There is a more obvious impact on the image-generating techniques, with both techniques remaining more or less constant up to about 2,048 points, and then slowing beyond that. However, the slowdown is not as dramatic as the text-generating techniques; even the largest data set only takes about nineteen seconds overall. Image overlay does display a slight advantage of about half a second over image generation for the largest data set, but further experiments are needed to determine whether this is statistically significant.

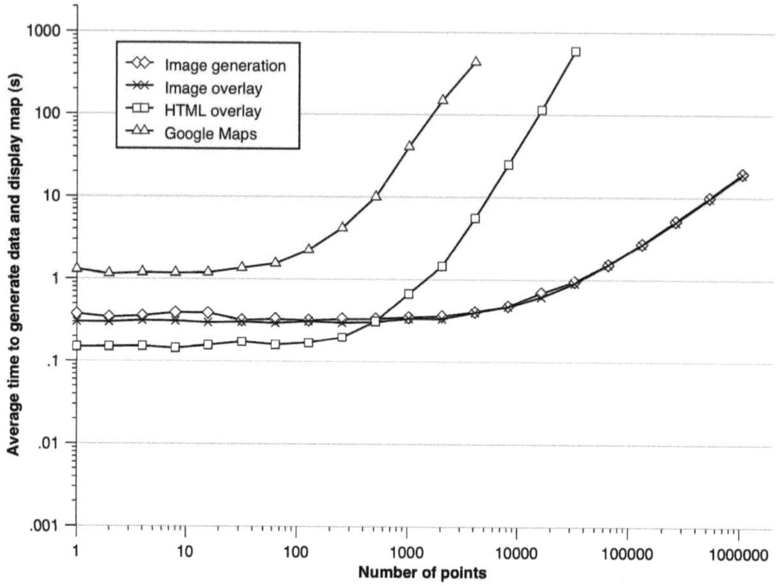

Fig. 9.10. Comparison of combined page load time for each technique (log-log scale)

5.3 Memory Consumption

Both the real and virtual memory consumption of the browser were measured before and after each test run. This provided the size of both the current "working set" and the total memory footprint of the browser process after it had completed a test run. The real memory results are shown in Fig. 9.11 and the virtual memory results in Fig. 9.12.

While both plots display similar trends, the real memory data are somewhat problematic. Real memory consumption was generally consistent across test runs, but would also frequently double for no readily apparent reason. This is particularly apparent with HTML overlay beyond 1,024 points. This was probably a consequence of other processes on the test machine requesting memory. The validity of the real memory data is thus somewhat doubtful, but they are broadly consistent with the virtual memory data. The virtual memory data proved more consistent overall, as the virtual memory footprint of a process is less likely to be impacted by other running processes.

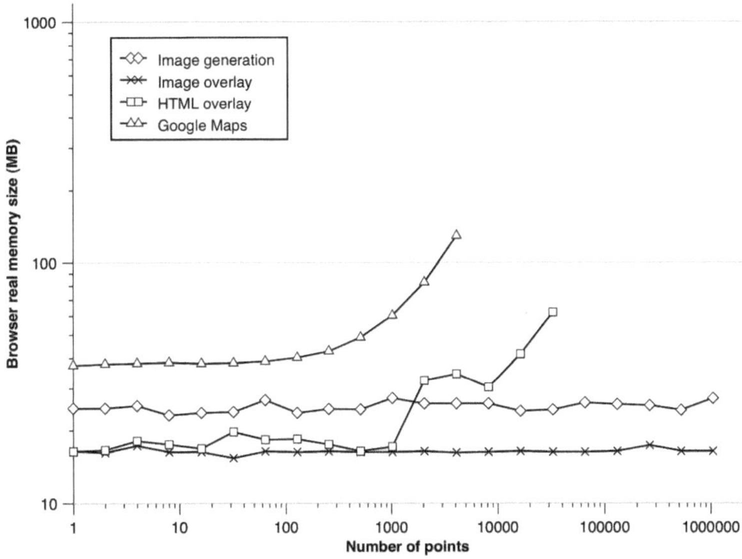

Fig. 9.11. Comparison of browser real memory consumption for each technique (log-log scale)

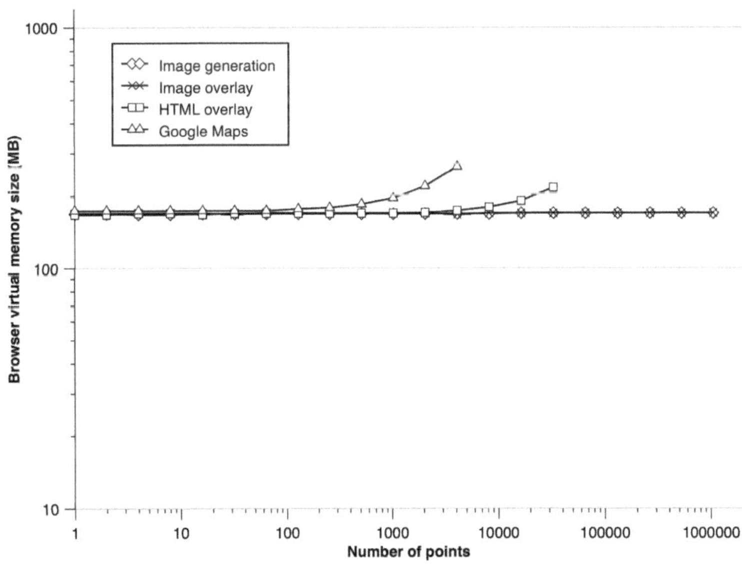

Fig. 9.12. Comparison of browser virtual memory consumption for each technique (log-log scale)

The results show that the two image-generating techniques have essentially constant virtual memory consumption of about 170 MB. This is to be expected, given that the generated data volume is also essentially constant. The text-generating techniques, however, clearly begin to diverge as the number of points increases. HTML overlay starts to visibly diverge somewhere between 2,048 and 4,096 points, reaching a maximum of about 216 MB when testing was terminated at 32,678 points. Google Maps starts to visibly diverge between 64 and 128 points, reaching a maximum of about 264 MB when testing was terminated at 4,096 points. This is in line with the initial expectation for these techniques that memory consumption would increase in proportion to the data set size.

6 Conclusions and Future Work

The scalability of four techniques—server-side image generation, server-side image overlay, server-side HTML overlay and distributed HTML overlay (Google Maps)—for online geographic visualization of Web site hits was tested. The results clearly show that server-side image generation and server-side image overlay scale the best from small to large data sets. Both HTML overlay techniques work well for small data sets, but their performance deteriorates rapidly with increasing data set size until they become unusable.

Despite this clear difference in scalability, there are still interesting questions remaining. The *model interaction environment* distribution style was not investigated, as it was unclear whether it could be achieved using only client-side JavaScript. This is an avenue for further investigation, and the appearance of native SVG support in browsers may be a viable option for implementing this distribution style. It would also be interesting to investigate the technologies rejected in this research (e.g., client-side Flash).

It was somewhat surprising that server-side HTML overlay and Google Maps were not consistent in where the different measures diverged, as shown in Table 9.2. Google Maps appears more consistent than server-side HTML overlay in this regard, but it seems logical to expect some correlation, so further investigation is required. Implementing an instrumented Web browser and server may enable more precise data to be gathered.

Shortly after completing the experiments, the author found the *msCross Web GIS client*[9], an open source Google Maps clone. Its documentation implies that it may be possible to build a self-contained implementation that would enable testing on an isolated network with client and server on

[9] http://datacrossing.crs4.it/en_Documentation_mscross.html

different machines. Network transfer time could then be measured, and issues arising from running the client and server on the same machine would be eliminated. This would require a distributed measurement infrastructure similar to that developed by Barford and Crovella (1999).

Table 9.2. Approximate number of points at which each measure begins to visibly diverge, for the server-side HTML overlay and Google Maps techniques

Technique	Data size	Map display time	Virtual memory
Server-side HTML overlay	64–128	128–256	2,048–4,096
Google Maps	256–512	64–128	64–128

The goal of this work was to identify the best technique for plotting the numerous downloads and abstract views from the Otago School of Business digital repository. The experiments were generic, however, so the results should be generalizable to any online visualization involving geographic data sets. Huge data sets are already commonplace and will only grow in future, so scalability of visualization techniques is a serious issue for all Web cartographers. What works for a small data set may not work for a large one!

The results clearly show that both server-side HTML overlay and Google Maps are inappropriate for large data sets. This leaves a choice between two similarly performing techniques: server-side image generation and server-side image overlay. However, multi-layer techniques display many practical advantages over single-layer techniques, such as the ability to dynamically toggle overlays, thus providing greater flexibility and a more dynamic user experience. Taking these benefits into consideration, the server-side image overlay technique is the clear winner.

Acknowledgements

The author would like to acknowledge Dr. Antoni Moore and Prof. George Benwell for their input into this research, and the reviewers for their valuable comments. Google and Google Maps are trademarks of Google Inc.

References

Andrienko G, Andrienko N, Dykes J, Mountain D, Noy P, Gahegan M, Roberts JC, Rodgers P, Theus M (2005) Creating instruments for ideation: Software approaches to geovisualization. In: Dykes J, MacEachren AM, Kraak J-M (eds) Exploring geovisualization. Elsevier, Amsterdam, pp 103–125

Babcock C (2007) Data, data, everywhere. InformationWeek, January 9, http://www.informationweek.com/story/showArticle.jhtml?articleID=175801775 (accessed on 24 May 2008)

Bar-Zeev A (2007) How Google Earth [really] works. RealityPrime blog, July 3, http://www.realityprime.com/articles/how-google-earth-really-works (accessed on 22 May 2008)

Barford P, Crovella M (1999) Measuring Web performance in the wide area. ACM SIGMETRICS Performance Evaluation Review 27(2):37–48

Bates PC (1995) Debugging heterogeneous distributed systems using event-based models of behavior. ACM Transactions on Computing Systems 13(1):1–31

Beaumont JR (1991) GIS and market analysis. In: Maguire DJ, Goodkind MF, Rhind DW (eds) Geographical information systems, vol 2: Applications. Longman, Harlow, UK, pp 139–151

Cammack RG (1999) New map design challenges: Interactive map products for the World Wide Web. In: Cartwright W, Peterson MP, Gartner G (eds) Multimedia cartography. Springer, Berlin, pp 155–172

Chang F, Dean J, Ghemawat S, Hsieh WC, Wallach DA, Burrows M, Chandra T, Fikes A, Gruber RE (2006) Bigtable: A distributed storage system for structured data. In: Proceedings of the 7th symposium on operating systems design and implementation (OSDI '06). Seattle, Washington, pp 205–218

CIA (2008) The world factbook. https://www.cia.gov/cia/publications/factbook/. Central Intelligence Agency, Washington DC (accessed on 24 May 2008)

Dent BD (1990) Cartography: Thematic map design, 2nd edn. Wm. C. Brown Publishers, Dubuque, Indiana

Dodge M, Kitchin R (2001) Mapping cyberspace. Routledge, London

Eick SG (2001) Visualizing online activity. Communications of the ACM 44(8):45–50

Enslow PH (1978) What is a "distributed" data processing system. IEEE Computting 1(1):13–21

Garrett JJ (2005) Ajax: A new approach to Web applications. Web essay, http://www.adaptivepath.com/publications/essays/archives/000385.php. Adaptive Path, LLC (accessed on 25 May 2008)

Gibbins H, Buyya R (2006) Gridscape II: A customisable and pluggable grid monitoring portal and its integration with Google Maps. Technical report GRIDS-TR-2006-8, May 12. Grid Computing and Distributed Systems Laboratory, The University of Melbourne, Melbourne, Australia

Golub E, Shneiderman B (2003) Dynamic query visualisations on World Wide Web clients: A DHTML solution for maps and scattergrams. International Journal of Web Engineering and Technology 1(1):63–78

Google (2008) Google Maps API. http://maps.google.com/apis/maps/ (accessed on 24 May 2008)

Jiang B, Ormeling F (2000) Mapping cyberspace: Visualizing, analysing and exploring virtual worlds. The Cartographic Journal 37(2):117–122

Kingston R (2002) Web-based PPGIS in the United Kingdom. In: Craig WJ, Harris TM, Weiner D (eds) Community participation and geographic information systems. Taylor & Francis, London, pp 101–112

Kitchin R, Dodge M (2002) 'There's no there there': Virtual reality, space and geographic visualization. In: Unwin D, Fisher P (eds) Virtual reality in geography. Taylor & Francis, London, pp 341–361

Lamm SE, Reed DA, Scullin WH (1996) Real-time geographic visualization of World Wide Web traffic. Computer Networks and ISDN Systems 28(7–11):1457–1468

Longley P, Goodchild M, Maguire D, Rhind D (2005) Geographic information systems and science, 2nd edn. John Wiley and Sons, Chichester, UK

MacEachren AM (1998) Cartography, GIS and the World Wide Web. Progress in Human Geography 22(4):575–585

Maxmind (2008) GeoLite City: Free geolocation database. http://www.maxmind.com/app/geolitecity. Maxmind LLC (accessed on 24 May 2008)

Offutt J (2002) Quality attributes of Web software applications. IEEE Software 19(2):25–32

Papadakakis N, Markatos EP, Papathanasiou AE (1998) Palantir: A visualization tool for the World Wide Web. In: Proceedings of the INET'98 conference. Geneva, Switzerland

Sale A, McGee C (2006) Tasmania statistics software. http://eprints.utas.edu.au/262/. University of Tasmania, Hobart, Australia (accessed on 25 May 2008)

Sayar A, Pierce M, Fox G (2006) Integrating AJAX approach into GIS visualization Web services. In: Proceedings of the advanced international conference on telecommunications and international conference on Internet and Web applications and services (AICT/ICIW 2006). IEEE Computer Society, Guadeloupe, French Caribbean, pp 169–175

Stanger N, McGregor G (2006) Hitting the ground running: Building New Zealand's first publicly available institutional repository. Discussion paper 2006/07. Department of Information Science, University of Otago, Dunedin, New Zealand

Stanger N, McGregor, G (2007) EPrints makes its mark. OCLC Systems & Services: International Digital Library Perspectives 23(2):133–141

Wood J, Brodlie K, Wright H (1996) Visualization over the World Wide Web and its application to environmental data. In: Yagel R, Nielson GM (eds) Proceedings of IEEE visualization '96. IEEE Computer Society and ACM, San Francisco, California, pp 81–86

Zhao H, Shneiderman B (2005) Colour-coded pixel-based highly interactive Web mapping for georeferenced data exploration. International Journal of Geographical Information Science 19(4):413–428

AFTERWORD

Developing Concepts for an Affective Atlas

William Cartwright[1], Adrian Miles[2], Brian Morris[2], Laurene Vaughan[2] and Jeremy Yuille[2]

[1]School of Mathematical and Geospatial Science, RMIT University, Melbourne, Australia

[2]School of Applied Communication, RMIT University, Melbourne, Australia

Abstract

Traditionally, cartographers have produced atlases as 'composite' products containing maps, photographs, diagrams and text. They are produced on paper, discrete media and distributed media. They are 'pre-composed' or 'constructed' products, whereby the cartographer controls all of the atlas content and how it is delivered to the user. It is a 'predictable' product, where the user is only a consumer and has no input into the design or content of the atlas.

Web 2.0 presents a new view on what can be done when provisioning users with cartographic materials. It is a different way of delivering cartographic media which, in many cases is basically non-cartographic, but delivers information that needs to be spatially defined and controlled if usable geographical information is to be assembled. It is a new publishing genre for cartography where users become part of Web-enabled collaborative publishing consortia. The tools and methods of delivery are different and they need to be explored, appreciated and applied.

This paper addresses these questions in the contexts of cartography and communications system design. The concept of an 'affective' atlas is proposed and its usefulness and effectiveness for better provisioning users with appropriate geospatial resources is being evaluated by a research team at RMIT University, Australia. It brings together researchers from Cartography and Applied Communication. The paper outlines theory being developed about how geospatial information might best be delivered using Web 2.0 and social software and its implications for the relationship of cartography to social, cultural, and narrative practice.

Keywords: interactive atlas, Web 2.0, collaboration, communications system design, integrated media.

1 Introduction

1.1 On Method

This paper has explored an experimental writing method. Each participant, each from varying disciplinary fields and practices, has been invited to respond to the idea of an 'affective atlas' and to discuss this from the point of view of its possibilities in relation to contemporary communication and cultural practice, their disciplinary assumptions and what an 'affective atlas' may provoke within our practice or disciplines. This has produced a paper that is, if you'll pardon the pun, a 'mapping' which is intended to serve as an introduction and probe into what we collectively believe to be a series of emerging possibilities and implications for cartography in a Web 2.0 enabled world.

1.2 Changing Landscapes

With the rapid proliferation of publications readily available through contemporary communications and multimedia publishing systems the way in which we access and produce information has changed forever. Maps and their digital underpinnings able to be downloaded, manipulated, distributed and republished a new genre of spatial artefact has now stabilised and become an accepted tool for exploring geography and mining geographical information. However, this has resulted in new ways to use these products and new ways of assembling data into personalised cartographic products and the subsequent proliferation of geospatial products that have been produced (in many instances) by ignoring the 'rules' that govern the integrity of good map compilation and design.

Where, previously, data and user were 'merged' by the provision of a particular mapping product that was generated to meet a certain usage requirement for viewing geographical information within a designated area contemporary products have changed the process. Users can become the map drawer, data can be assembled from many discrete and geographically dispersed sites and visualization products can be generated using a plethora of depiction techniques. This allows data to be interpreted into usable maps using software that is readily available and inexpensive. Everything has changed.

Traditionally, geographical visualization products have been developed to provide information about 'real' geography. But with the plethora of devices and media resources now available users can also access information that is provided differently than how it is provided by the 'mapping' world. They can use popular media (like books, movies and videos) to experience a view of geography that is presented in an altogether different way than conventional geographical visualization tools offer. In many cases these resources include depictions of different or synthetic realties, as well as our 'real' world, which the viewer/user must accept to enjoy or be involved with what is presented.

1.3 Visualization Artefacts and Distortion

The basic premise of using maps and other (geo)visualization artefacts is to build mental models of reality. This evokes the image of 'being there', without (actually) being there. Historically, maps have been used to provide information to users about places recently discovered or voyages completed to unknown worlds or hitherto seemingly impossible journeys. But, cartographers and map publishers have only been able to 'map' available information. When geographical knowledge about the world was limited the depiction of the world was also limited.

But the problem was what was to be done about the blank spaces? Some of the earliest recorded maps leave blanks or spaces of nothingness to reflect unknown lands and locations, factors that aren't yet known. An example of this is the ancient maps of the Babylonians who integrated both the known and the unknown within their maps of their world. 'What the Babylonians knew, their property lines and city walls, they mapped with some attention to accuracy: what they did not know, the lands beyond their own, they chose either to ignore or to fabricate' (Noble Wilford 2002, p 11).

As geographical knowledge expanded, so did the contents of the map. The depiction of the world demanded information, and if information was not available it was either left out, or icons depicting geographical and social scenes or events (in many cases not from the region in question at all) were substituted for the usual blank spaces. The image was distorted so as to provide a more complete 'picture', but an incorrect reality was mapped. This phenomenon has not really changed when contemporary techniques are used, for example. When user/producer maps only realize information about their particular 'place', rather than 'space'.

In order to 'be there' we generally rely on maps to find a location, or to navigate to a place of interest. But, as a methodology for constructing our individual worlds (our mental model of reality) mapping can be messy. There can be times of great confusion before we find clarity, and as such it calls for an awareness of the subtleties and possibilities of mapping. In constructing a map we endeavour to make sense of the components or elements and we look for features of meaning and significance so that we can relate them to each other in space. We use the map and the mapping process to both conceive and capture the world around us and thereby bring it into a tangible or 'manageable' form. Maps reflect what we know and what we don't (Lippard 1998). By tracking the evolution of maps we are able to see the evolution of our knowledge and our journey as individuals and as a society.

2 Engaging in Cartography as Everyday Practice

In 'The Practice of Everyday Life', Michel de Certeau (1984) explores the concept of practice as a means of engaging with the activities of life. To practice is to partake in the actions of life; it is a mode of engagement. De Certeau argues that to understand the practices of everyday living, is to engage with our distinctly individualistic ways of knowing. This is an argument based on the premise that knowledge comes from, and draws upon, our personal understandings in order to have meaning for us. He gives the example that to gain knowledge and interact with the external texts of our lives it is necessary for us (as individuals) to personalise, customise or mutate that knowledge into a form that has meaning for us, and through this we inhabit the space of the text. This is a form of inhabitation that could be understood as a mutation, a modification of the external and untouched world outside of us.

'The mutation makes text habitable, like a rented apartment. It transforms another person's property into a space borrowed for a moment by a transient. Renters make comparable changes in an apartment they furnish with their acts and memories. In the same way the users of social codes turn them into metaphors and ellipses of their own quests.' (de Certeau 1984, pp xxi - xxii)

In this way each of us, in our search for meaning is like the renter. We inhabit the space of the practices of everyday life; in creating them we navigate, negotiate and construct this space of habitation through our knowledge, memories and expertise.

To navigate, negotiate and construct spaces and to transform these actions into something that has meaning for us we need to be able to 'tag' our acts and memories (from the past) and our proposed future acts in projected spaces with not just 'geo' elements ("you were here doing this", "you will be there doing that"), but we need to personalize this space by adding other tags. The spaces become more personal, the artefacts we use to make spaces 'ours' become more affective. But, to do this – to personalise spaces using affective artefacts – requires a tool. We believe that this tool can be the assimilated and connected artefacts that can be delivered using Web 2.0. And, if these artefacts are both personal and geo-coded, they might offer the potential for users of traditional (rented) cartographic artefacts to actually 'own' that information.

3 Back to the Future: Web 2.0 and Cartography

3.1 The Web and Mapping

The Web provides mapping products that are now desktop-delivered, provided to the mobile user and served in ubiquitous installations. They are delivered via the telecommunications conduit of large telecommunications companies as proprietary product or via Open formats. Contemporary communications and (geo)information methods provides tools for constructing personalized, or affective geospatial products. These are:

1. Personal and global electronic publishing; and
2. Collaborative global electronic publishing and communication.

The combination of these two (geo)information provision developments has caused cartographers to re-think how some maps should be composed, delivered and used. This 'New Cartography' uses Web 2.0 ideas as the basis for a different, collaborative form of communication.

3.2 Web 2.0 and Social Software

The confluence of relatively ubiquitous bandwidth (in the west at least), high CPU speeds and adoption of web standards has led to the emergence of a second generation of web based services. Dubbed "Web 2.0" by O'Reilly Media, the concept was originally touted in Tim O'Reilly's "Open Source Paradigm Shift" (2004). This paradigm shift has changed the way in which people access and use information on the Web, including geospatial information. But, to work, it needs people, and people who are willing to collaborate and share information construction and delivery. To do this information and software needs to be shared and products, like mapping products, built in entirely different ways, whereby 'the map' may be composed and provided from many, Web-connected, resources, and not just one, as was traditionally the case.

In this article, O'Reilly uses the long-term trends of software commoditization, network enabled collaboration and software customizability to describe a next generation networked services exhibiting the following characteristics:

- Services (in this context, maps, images, annotations, text, etc. – 'things' that we would expect to appear on maps and atlases) are delivered via the web, as a web site (that is, the product does not exist at all before the 'construction' of that product is completed using many resources;
- The web is used as a "platform", and services open Application Programming Interfaces (API's) for others to extend their services (for example, an existing base map might be accessed from one site, annotated on another with symbols, and images added from another site, as per mapping sites like *Google Maps*® and *Google Earth*®); and
- Services are designed to work in a social manner - they need people (not just one cartographer per (generated) map, but many contributors) – e.g.: a recommendation service is useless without any members, may be satisfactory with a few hundred members, but is outstanding with a few million.

O'Reilly (2004) coined a term "architecture of participation" to describe the nature of systems that are designed to encourage user contribution.

In hothouse fashion, the explosion of Web 2.0 services led to further standardization, and microformats. One side effect of so many different applications publishing API's is that it is now relatively easy to mix two or more services together to create new representations of previously unrelated information. One example is mixing rental information form a site like craigslist.com with a mapping service like Google maps, to give a geographic representation of available rental properties (for this example see http://www.housingmaps.com/). This method of mixing services has earned the name "mash-up".

Some examples of Web 2.0 services and applications include:

- Blogging sites, and open source software like *WordPress* or *Drupa*l;
- *Wikipedia* and other instances of user editable websites (or wiki's);
- Customer driven suggestion, like Amazon's book suggestions, or Google's *AdSense*;
- Social networking sites like *friendster*, *linkedup* and *myspace*; and
- Mashups such as *ning* video.

4 Maps of Place into Maps of Space

4.1 Mental Constructions of the Universe

Early civilizations believed that they were at the centre of everything and, compared to contemporary communities, they were also relatively isolated. The artefacts they produced generally reflected this viewpoint. Generally, according to Berthon and Robinson (1991) humans then were looking for responses to two basic questions:

- Where am I? and
- What happens to me when I die?

Mental constructions of the universe were devised to deal with these two mysteries, and creating their own pictures of the world allowed their imagination of what the world was like to be rendered in imagery. Humankind has always desired to construct an image of the world they knew (or a world that they thought existed). These artifacts recorded the geographical extent of their knowledge, usually restricted by things like mobility or religious beliefs, and they portrayed the elements of the world deemed important or thought necessary to 'build' a mental map of their world. The extent of these maps, what they showed and how information was illustrated was dictated by their personal, mental and cultural construct about their world.

4.2 Maps as Cultural Practice

A map is a cultural technology. More recent cultural studies perspectives have argued that a technology is not autonomous (as it is often assumed in common sense understandings), but is instead 'integrally connected to the context within which it is developed and used; that culture is made up of such connections; and that technologies arise within these connections as part of them and as effective within them' (Slack and Wise 2005, p 112). So cartography and maps themselves might be understood not simply as singular and discrete representational techniques and technologies, but as an articulated technology, constituted by dynamic assemblages of 'practices, representations, experiences, and affects' (Slack and Wise 2005, p 129).

A common trait of institutionally-produced maps, however, is that they do make claims regarding their authoritative uniqueness (e.g., through their level of accuracy due to the detail of data they have accumulated; or, perhaps, their innovation—and therefore increased usability—due to their style of representation of that data). These claims to distinction reinforce a sense of that particular map's autonomy and abstraction from the social world (and its eventual use). However, as noted by Miller (2006, p 197), Web 2.0 mapping and GIS products might provide the answer:

"The point is that the Google Maps mashups model has put into public practice the notion that an accessible, agile, adaptable GIS can be built that accepts direct, local, even vernacular public input and, in turn, puts out usable, unique, localized, and important results."

A paradox emerges via this technological innovation. On the one hand, maps and their information technology descendants such as databases appear to clearly manifest a historical desire for a totalizing view and knowledge of the world. On the other hand, the more the technological ability to network and combine these extensive databases of data grows, the more—it seems—we become aware, as argued by Latour and Hermant (2006), that many of these different 'forms of reference can coexist without ever being entangled' (p 37).

5 Augmented Place

Placing ourselves within the World, be it our world or the 'worlds' of others, requires us to make sense of that world, to inhabit it through connections, that are read and communicated. We position ourselves in relation to the various characteristics and possibilities of the location we are in, and we imagine connections to other places located somewhere else. However as argued by Weinberger (2002) on the Web we are able to 'experience something we can never experience in the real world: places without space. Instead of needing a containing space to enable movement, the Web has hyperlinks. Links are at the heart of the Web and the Web's spatiality'. It is links, and our ability to navigate and construct them, which enable us to create locations on the Web; to place ourselves in multiple locations without the limitations of space. This is what makes the Web 'place-ial' (Weinberger 2002) a series of interlinked places not limited or confined by the challenges or realities of travel. We can traverse between places and create our own connections and interconnections between multiple sites without being limited by physical or temporal constraints.

Web 2.0 technologies continue an ongoing process of the steady virtual augmentation of place via various media forms and content. In their studies of the location-aware, Japanese mobile phone-based game of *Moji*, Licoppe and Inada (2006) direct us to some of the potential effects and affects of such technologies:

- Users 'adjust their displacements to the 'augmented' ecologies that they encounter' – so the 'new' content produced through these collaborative maps will contribute to individuals recasting their movements and engagements with the re-presented environments they inhabit.
- These technologies potentially recast the dynamics of encounters between strangers (which has been vital to understandings of civility and interaction in the public sphere) for they may involve not simply the inhabitation of one's own mediated world but refigured forms of interaction (eg, through 'chat'/IM functions built into location-aware devices).

As they argue: "[Advanced mobile services and devices] participate in a real engineering of encounters between people and things, in both material and immaterial forms. They are set to play a key part in determining the way in which information and communication technologies reshape our structures of anticipation, that is, our perceptions and expectations concerning the ways in which the entities constituting our environment can act and appear to us, here and now" (Licoppe and Inada 2006).

Web 2.0 won't of course require cartographers to abandon the techniques and standards engendered by their professional and institutional training (though it might open up expanded possibilities of what it means to be a cartographer and a rethinking of the objects that those techniques are applied to). These new maps will potentially become part of newly assembled and visible private-public geography. These maps connect and make visible the hitherto present but unrepresented frames that organise an individual's sense of place. New Web 2.0 technologies (e.g., allow significant additions, expansions and assemblages of data like tagging of locations; incorporation of locational data etc, and mobile articulations of multiple frames of place.

6 Maps and Affect

A map, well at least pragmatically, can be thought of as a particular sort of instrumental image. It is instrumental because maps are (setting aside for now the obvious case of artist's maps), intended to be cognitive and empirical tools that have some purchase and actionable reference to those things the map claims to map. They are images because map making is the practice of developing and using graphic cues and symbols as indices of that which is mapped. In this manner a 'good' map would be that map most instrumentalises — makes transparent, apparent and plain — this relation, and so makes the map the most accurate according to these empirical criteria. In contemporary cartography this instrumentalisation (which has always driven and informed map making, after all an inaccurate map is not so much useless as just plain dangerous) has probably been expressed along two key lines of thinking.

One line has been the development of ever richer and more complex maps and mapping projects — aided by developments in computer visualation, GPS, and improved measurement devices and systems. The second line is speed and describes the manner in which real time mapping is now undertaken, for example through live GPS data collection, the distribution of GPS with subscription map services and the real time collection of cartographic data locally and remotely, all of which can be immediately published as maps via real time services. (As an aside, each of these axes has also promoted the embedding of maps into the everyday, evidenced via such popular projects as *Google Maps®*, *Google Earth®,* hand held GPS, car navigation systems and location aware electronic devices and services.)

However, how we use maps, that is not how cartographers make maps but how map users apply maps, produce affective maps. That is, that actual way in which they use maps, annotate them, insert personal insights of their own interpretations of geography makes the map affective. In this conception affect is understood to be an abstract remainder which cannot be expressed or spent by just using a map as a locational or navigational instrument. I can try to 'use' it, but in using it something is left and this left over thing produces, and is, affect. For example, my child receives numerous instrumental actions of care, nurturing, tendering, parenting and so on, but each of these actions, even in total, is never adequate to the desire and wish to do these actions. This remainder, which is not exhausted through all these applied actions, is affect and in this case we could call it paternal or parental love. Alternatively, when I use a map to find my way along a bush track, then discover a fine swimming hole which I then pencil in on my map the map itself is no longer only instrumental — even if I use it to re–find this swimming hole in the future. It now has connotations of discovery, pleasure, water, wetness and so on. The point of affect is that it is what remains after action, and while lots of actions don't produce affect (I pick up my pencil, my heart beats) many do and this remainder is the excess that action does not resolve or expend.

Now, a 'good' map, as we saw, has no such excess, it intends to map its territory with as much empirical verisimilitude as possible. In other words the use of the map cartographically is intended to exhaust and expend itself utterly. Hence, ambiguity is avoided and standardised empirical data provides the bedrock for a map's ability to address its territory. Maps, except perhaps in the daydreams of cartographers, never reach this ideal point, but this remains that which they aspire to. However, when maps are used they soften and become affective through this everyday use. We draw on them, tracing our route through a strange city, or perhaps simply mark in the location of a favourite café. The point now is not only the versimilitude of the map to its world but also its indexical relation to the personal. My map now has the coffee stains of the journey itself. The annotations <u>on</u> the map make it affective not just for the initial user of the map, but for other users as well. As a user we can read our own memories into the map image, remembering a walk along a train or a serendipitous journey through a hitherto unknown town. As individuals we can use our own mental map to make the physical map affective. But, to make the map affective for other users the map needs to be annotated.

Now, with these axes of the improving quality and granularity of data, an increasing speed in the production of the 'map', and their widespread informal use via emerging digital technologies, the possibility arises for maps that are, for want of a better term, less instrumental. It is now possible to build maps, and of course atlases, that utilise digital and hypertextual technologies to allow our 'coffee stains' to inform the map, to be shared with others, and to become a part of cartographic practice. For example, my personal GPS device may automatically trace my route through a national park, any photos I take are dynamically added, with appropriate cartographic metadata, and I add (or have collected) further social metadata such as my age, gender, whether I appear to be in a group, if my journey mirrors others, and so on. This new map shifts from being instrumental and distinct from that which is mapped to being affective, a site for stories and other emotional trails, a palimpsest of my and others experiences recorded within and by a different sort of map — but a map nonetheless.

Such an affective atlas creates a gap between the instrumental application of cartography and the pragmatics of map use, and introduces a possible cartographic practice that blurs the boundaries between cartographer, mapping, and the map user. The trick is not in working out how to let such maps be annotated, but how such annotations become materials for mapping.

7 'The Academy' vs 'The Innovationist'

We argue that developers of mapping products using Web 2.0 do provide a more innovative approach to the portrayal of geographic information. However, the basics of providing access to applicable mapping artefacts are still being led or dictated by 'The Academic Rules' of cartography. That is, the means of data presentation and access is really no different from that used in conventional map production. However, leaving aside the actual map production process, and focusing on the facilitation of a system that includes all steps from map concept to the realization of a final product, Web 2.0 and associated social software demands that the cartographic thought process be re-visited.

Innovation sometimes means cutting loose the traditional ties that bind the cartographer to conventional methods of information communication. The need to use methods that remove the cartographer from the delivery component of the package may not be a high priority, as, from the Academy perspective, the professional 'packager' of geo-spatial information needs to always be in command of the delivery. of those packages. Innovative tools now empower users to embark on real discovery and enlightenment projects to better inform about things geographical and for the geographical to inform the world.

Certain academy or traditional rules govern the way in which data is collected and presented. It can also govern the way in which a geographical information package is used. Particular users could access and use geographical information in different ways, rather than only by being passive users of geographical information products and packages provided by 'The Academy'. That is, if 'individualised' interactive packages were self-built, or built in a collaborative manner (cartographer and 'production tool-literate' user), they develop different, personalised packages with which to view and interpret geographical information products from different, non–specialist perspectives that in turn can be, and are, made available to other users. Are such maps better than those from the academy? Using annotations, they can be built atop of maps provided by the academy – providing academic rigour with personalized annotations of a journey through space. We do not contend that these personal, annotated maps are superior to maps from the academy (we have no proof of this), but we propose that the **use** of such maps needs to be evaluated and their effectiveness ascertained.

8 Future Research

Social Network Analysis (SNA), small world networks and the emergence of social software facilitated collaborative environments that all raise significant possibilities for cartography. These systems rely on what are described as 'weak' links (links that loosely tie-together the elements and portray the connections between the elements mapped (perhaps what Bertin (1983) would call 'correspondences') to facilitate the development of rapid and short route connections between otherwise disparate individuals, locations and ideas. Such systems are in themselves maps, however they are emergent, transitory, and unstable. In addition they provide the communicational and technological glue to allow new forms of information and knowledge to be developed, expressed and exchanged. In this economy knowledge is no longer something that is reasonably solid and then represented in a 'package' but can evolve and arise within practice itself. This field, which is enormous and deeply interdisciplinary, provides much of the basis for future research within the realm of an 'affective atlas'.

This research will address the following questions:

- What is an affective atlas?
- What do we produce when making such atlases?
- Should we still only compose and publish atlases in the conventional manner?
- Should we consider collaborating with users to publish atlases as user/producer 'partnerships'?
- Can 'amateur' content be included in professional cartographic practice and products?
- What changes to cartographic practice do these technologies require (or provoke)?
- What does cultural studies and related humanities disciplines offer this?
- How does communication design intersect with the design, description and critique of the affective atlas?
- What happens to cartographic rigour when non-cartographers can add data?

9 Pilot Project

The first Pilot project will be developed in conjunction with Parks Victoria, Australia. Parks Victoria is the custodian of a diverse estate of significant parks in Victoria and of the recreational management of Port Phillip Bay, Western Port and the Yarra and Maribyrnong rivers. It manages parks and reserves of approximately 3.96 million hectares (17% of the State of Victoria). Through effective environmental and visitor management, Parks Victoria is dedicated to preserving the natural and heritage values of parks, bays, and waterways. As well, the organisation has plans to develop a Parks Victoria Centre of Excellence, to be located at Wilsons Promontory Marine National Park, the largest Marine National Park in Victoria, which spans 70 kilometres of mainland coastline and covers 15,550 hectares.

Parks Victoria currently holds extensive information sets and materials relating to its national parks. The information is stored as texts, diagrams, reports, photographs and maps. There are concerns that these data sets, although actually available, it is not feasible to have them at-hand when and where needed to support decision-making. Parks Victoria has already developed large digital repositories of geospatial data for its areas of responsibility and has the desire to further exploit its existing systems in innovative ways. Parks Victoria is interested in developing effective ways in which to make this data available to all stakeholders in effective and sustainable parks management and operation.

Building and evaluating integrated media data repositories / information access tools are of significant interest to Parks Victoria. In particular Parks Victoria already supports its park management by exploiting spatial information through the use of GIS. A system that provided a system for data collection, storage, analysis and access to information that was defined by the location of that information, and a system that was built in a collaborative manner using contemporary Web 2.0 tools would support more efficient management of their parks.

The project is based on the following premises:

- The production of media artefacts has moved from being a specialised, elite and broadcast practice to one that is everyday, personal, and distributed via the Web.
- Media artefacts, through convergence, are no longer discrete. This happens at the level of the devices we use to gather media (where a mobile phone is a still and video camera, audio recorder, telephone, web browser and recorder of live geospatial data), and at the level at which we collate and view data (the computer and the Web).

- The relations or connections that can be identified between parts (text, video, sound, maps, scientific reports, photographs) can be as significant as the parts themselves; therefore
- Information management has moved from the maintenance of archives and related material to systems that facilitate the discovery or emergence of knowledge from within archives and the enhancement of knowledge systems through the appropriate collection, classification and publication of new artefacts in archives.

The project will develop integrated media software that will complement existing software. It will allow access to the myriad of existing parks management data and records, both hard-copy and digital, by building a repository that can be appended, updated and interrogated in multiple locations, when mobile and in fieldwork and data collection situations. Specific theoretical strategies will be developed to guide software development and then to apply these tools to parks management.

10 Conclusions - Application of Computers and a Different Production Model

Cartography needs to be re-visited in the light of the application of social software technologies. Is cartography any different when composed and delivered using social software? Does cartography need to be re-defined because of the revolution that has taken place within the ways in which information is now communicated, the types of information that can be transferred, the scale of this afforded by contemporary communication technologies, and how cartographic and geospatial information and knowledge is now being produced globally?

Considering the tremendous impact that information technology has had on the graphic arts in particular, and also on the possibilities for producing fairly professional products by non-cartographers, the area of responsibility for cartographers perhaps needs to be re-defined as well.

New technologies enable non-cartographers to produce viable maps, which while viewed as naïve mapping products in the eyes of cartographers, are usable products (whilst perhaps inefficient, maybe scientifically inaccurate and artistically inelegant from a cartographer's viewpoint). These can be developed and produced without a cartographer's input whatsoever. Does cartography therefore need to be re-defined in terms of cartographer and also in terms of naïve producer/consumer as well? Considering that cartographers can control most elements of the provision of products until the final consumption of the product, perhaps a division needs to be made between the actual 'behind the scenes' elements of contemporary cartography and the 'public face' of cartography - 'consumer cartography'.

References

Bertin J (1983) The semiology of graphics. Univ. Wisconsin Press, Madison, Wisc

Berthon S, Robinson A (1991) The Shape of the World. Rand McNally & Company

Debord G (1981) Introduction to a Critique of Urban Geography, Situationist Anthology. K Knabb, Bureau of Public Secrets, Berkeley, pp 5-8

de Certeau M (1984) The Practice of Everyday Life. The University of California Press, Berkeley

Latour B, Hermant E (2006) Paris: Invisible City. http://www.bruno-latour.fr/virtual/index.html (accessed 12 February 2006)

Licoppe C, Inada Y (2006) Emergent uses of a multiplayer location-aware mobile game: The interactional consequences of mediated encounters. Mobilities 1(1):39-61

Lippard L (1998) The Lure of the Local. New Press, USA

Miller CC (2006) A Beast in the Field: The Google Maps Mashup as GIS/2. Cartographica 41(3):187-199

Noble Wilford J (2002) The Mapmakers. Pimlico, London

O'Reilly T (2004) The Open Source Paradigm Shift. http://tim.oreilly.com/articles/paradigmshift_0504.html (accessed 18 December 2006)

Slack JD, Wise JM (2005) Culture + Technology: A Primer. Peter Lang, New York

Weinberger D (2002), Small Pieces Loosely Joined, 1st edn. Perseus Books Group, http://www.smallpieces.com/content/chapter2.html (accessed 26 March 2002)

MIX
Papier aus verantwortungsvollen Quellen
Paper from responsible sources
FSC® C105338
FSC
www.fsc.org

If you have any concerns about our products,
you can contact us on
ProductSafety@springernature.com

In case Publisher is established outside the EU,
the EU authorized representative is:
Springer Nature Customer Service Center GmbH
Europaplatz 3, 69115 Heidelberg, Germany

Printed by Libri Plureos GmbH
in Hamburg, Germany